CHEMISTRY DEPT.
Vermont Tech College

# The Chemical Formulary

*A Collection of Commerical Formulas, Collected During 1968, for Making Thousands of Products in Many Fields*

**VOLUME XV**

*Editor-in-Chief*
H. BENNETT, F.A.I.C.
*Director, B. R. Laboratory*
*(Formula Consultants)*
*Miami Beach, Florida,* 33140

CHEMICAL PUBLISHING COMPANY, INC.
New York 1970

PRINTED IN THE UNITED STATES OF AMERICA

## Contributors

| | |
|---|---|
| Bowman, C. E. | Consultant |
| Carman, W. E. | Amer. Maize Prod. Co. |
| Cohen, R. S. | Dover Chemical Corp. |
| Dilworth, P. | Fritzsche Bros. |
| Feustel, W. K. | Vanderbilt Co., R. T. |
| Forwalter, J. | Armour Ind. Chemical Co. |
| Garizio, J. E. | Reheis Chemical Co. |
| Goldschmiedt H. | Mem Co. |
| Kennedy, J. N. | Camilla Hall Infiroary |
| Krepela, R. T. | Consultant |
| Levitt, B. | Consultant |
| Markoff, M. K. | Spencer Kellogg Co. |
| Mecca, S. B. | Schuylkill Chemical Co. |
| Mendell, E. | Mendell Co., E. |
| Morel, T. | Scher Bros Inc. |
| Nichols, M. J. | Goodrich Gulf Chemical Co. |
| Patureau, A. M. | Pennsylvania Ind. Chem. Corp. |
| Phares, C. | Goldschmidt Chemical Co. |
| Rosenthal, M. L. | Robeco Chemicals Inc. |
| Schoenberg, T. G. | Richardson Co. |
| Schumacher, G. E. | University of Toledo |
| Sheers E. H. | Arizona Chemical Co. |
| Skinner, G. W. | Bareco Division |
| Snyder, M. | Onyx Chemical Co. |
| Spindel, S. | Litter Laboratories, D. |
| Steele, F. J. | Ephrata Community Hospital |
| Szanto, J. | Consultant |
| Treibl, H. G. | Cloroben Chemical Corp. |
| Whatley, A. | Hampshire Chemical Div. |
| Whitener, P. D. | Winthrop College |
| Wolf, R. F. | Consultant |

# PREFACE TO VOLUME XV

This volume contains formulas collected in 1968. The only repetitious formulas are these in the Introduction (Chapter 1) which are given for those who have no technical training and experience in compounding. These will serve as a guide for beginners and students. They should read the introduction carefully and even make a few preparations before attempting more complicated formulas that follow.

H. BENNETT

NOTE: All the formulas in Volumes I to XV (except in the introduction) are different. Thus, if you do not find what you want in this voume, you may find it in one of the others.

NOTE: This book is the result of cooperation of many chemists and engineers who have given freely of their time and knowledge. It is their business to act as consultants and to give advice on technical matters for a fee. As publishers, we do not maintain a laboratory or consulting service to compete with them. Therefore, please do not ask us for advice or opinions, but confer with a chemist.

Formulas for which patent numbers are listed can be manufactured only after obtaining a license from the patentees.

# PREFACE TO VOLUME XV

# PREFACE

Chemistry, as taught in our schools and colleges, concerns chiefly synthesis, analysis, and engineering — and properly so. It is part of the right foundation for the education of the chemist.

Many a chemist entering an industry soon finds that most of the products manufactured by his concern are not synthetic or definite chemical compounds, but are mixtures, blends, or highly complex compounds of which he knows little or nothing. The literature in this field, if any, may be meager, scattered, or obsolete.

Even chemists with years of experience in one or more industries spend considerable time and effort in acquainting themselves with any new field which they may enter. Consulting chemists similarly have to solve problems brought to them from industries foreign to them. There was a definite need for an up-to-date compilation of formulae for chemical compounding and treatment. Since the fields to be covered are many and varied, an editorial board of chemists and engineers engaged in many industries was formed.

Many publications, laboratories, manufacturing firms, and individuals have been consulted to obtain the latest and best information. It is felt that the formulae given in this volume will save chemists and allied workers much time and effort.

Manufacturers and sellers of chemicals will find, in these formulae, new uses for their products. Nonchemical executives, professional men, and interested laymen will make through this volume a "speaking acquaintance" with products which they may be using, trying, or selling.

It often happens that two individuals using the same ingredients in the same formula get different results. This may be due to slight deviations in the raw materials or unfamiliarity with the intricacies of a new technique. Accordingly, repeated experiments may be necessary to get the best results. Although many of the formulae given are being used commercially, many have been taken from the literature and may be subject to various errors and omissions. This should be taken into consideration. Wherever possible, it is advisable to consult with other chemists or technical workers regarding commercial production. This will save time and money and help avoid trouble.

A formula will seldom give exactly the results which one requires. Formulae are useful as starting points from which to work out one's own ideas. Also, formulae very often give us ideas which may help us in our specific problems. In a compilation of this kind, errors of omission, commission, and printing may occur. I shall be glad to receive any constructive criticism.

H. BENNETT

## CONTENTS

# ABBREVIATIONS

| | |
|---|---|
| amp | ampere |
| amp /dm$^2$ | amperes per square decimeter |
| amp /sq ft | amperes per square foot |
| anhydr | anhydrous |
| avoir | avoirdupois |
| bbl | barrel |
| Bé | Baumé |
| B.P. | boiling point |
| °C | degrees Centigrade |
| cc | cubic centimeter |
| c d | current density |
| cm | centimeter |
| cm$^3$ | cubic centimeter |
| conc | concentrated |
| c.p. | chemically pure |
| cp | centipoise |
| cu ft | cubic foot |
| cu in. | cubic inch |
| cwt | hundredweight |
| d | density |
| dil | dilute |
| dm | decimeter |
| dm$^2$ | square decimeter |
| dr | dram |
| E | Engler |
| °F | degrees Fahrenheit |
| f f c | free from chlorine |
| f f p a | free from prussic acid |
| fl dr | fluid dram |
| fl oz | fluid ounce |
| fl pt | flash point |
| F.P. | freezing point |
| ft | foot |
| ft$^2$ | square foot |
| g | gram |

# ABBREVIATIONS

gal. . . . . . . . . . . . . . . . . gallon
gr. . . . . . . . . . . . . . . . . grain
hl. . . . . . . . . . . . . . . . . hectoliter
hr. . . . . . . . . . . . . . . . hour
in. . . . . . . . . . . . . . . . . inch
kg. . . . . . . . . . . . . . . . . kilogram
l. . . . . . . . . . . . . . . . . . liter
lb. . . . . . . . . . . . . . . . . pound
liq. . . . . . . . . . . . . . . . . liquid
m. . . . . . . . . . . . . . . . . meter
min. . . . . . . . . . . . . . . . minim, minute
ml. . . . . . . . . . . . . . . . . milliliter (cubic centimeter)
mm. . . . . . . . . . . . . . . . millimeter
M.P. . . . . . . . . . . . . . . . melting point
N. . . . . . . . . . . . . . . . . Normal
N.F. . . . . . . . . . . . . . . . National Formulary
oz. . . . . . . . . . . . . . . . . ounce
*p*H. . . . . . . . . . . . . . . . hydrogen-ion concentration
p.p.m. . . . . . . . . . . . . . . parts per million
pt. . . . . . . . . . . . . . . . . pint
pwt. . . . . . . . . . . . . . . . pennyweight
q.s. . . . . . . . . . . . . . . . . a quantity sufficient to make
qt. . . . . . . . . . . . . . . . . quart
r.p.m. . . . . . . . . . . . . . . revolutions per minute
sec. . . . . . . . . . . . . . . . second
sp. . . . . . . . . . . . . . . . . spirits
Sp. Gr. . . . . . . . . . . . . . specific gravity
sq. dm. . . . . . . . . . . . . . square decimeter
tech. . . . . . . . . . . . . . . . technical
tinc. . . . . . . . . . . . . . . . tincture
tr. . . . . . . . . . . . . . . . . tincture
Tw. . . . . . . . . . . . . . . . Twaddell
U.S.P. . . . . . . . . . . . . . United States Pharmacopeia
v. . . . . . . . . . . . . . . . . volt
visc. . . . . . . . . . . . . . . . viscosity
vol. . . . . . . . . . . . . . . . volume
wt. . . . . . . . . . . . . . . . . weight

Chapter I

# INTRODUCTION

The following introductory matter has been included at the suggestion of teachers of chemistry and home economics.

This section will enable anyone, with or without technical education or experience, to start making simple products without any complicated or expensive machinery. For commercial production, however, suitable equipment is necessary.

Chemical specialties are composed of pigments, gums, resins, solvents, oils, greases, fats, waxes, emulsifying agents, dyestuffs, perfumes, water, and chemicals of great diversity. To compound certain of these with some of the others requires definite and well-studied procedures, any departure from which will inevitably result in failure. The steps for successful compounding are given with the formulae. Follow them rigorously. If the directions require that (*a*) is added to (*b*), carry this out literally, and do not reverse the order. The preparation of an emulsion is often quite as tricky as the making of mayonnaise. In making mayonnaise, you add the oil to the egg, slowly, with constant and even stirring. If you do it correctly, you get mayonnaise. If you depart from any of these details: If you add the egg to the oil, or pour the oil in too quickly, or fail to stir regularly, the result is a complete disappointment. The same disappointment may be expected if the prescribed procedure of any other formulation is violated.

The point next in importance is the scrupulous use of the proper ingredients. Substitutions are sure to result in inferior quality, if not in complete failure. Use what the formula calls for. If a cheaper product is desired, do not prepare it by substituting a cheaper ingredient for the one prescribed: use a different formula. Not infrequently, a formula will call for an ingredient which is difficult to obtain. In such cases, either reject the formula or substitute a similar substance only after a preliminary experiment demonstrates its usability. There is a limit to which this rule may reasonably be extended. In some cases, substitution of an equivalent ingredient may be made legitimately. For example, when the formula calls for white wax (beeswax), yellow wax can be used, if the color of the finished product is a matter of secondary importance. Yellow beeswax can often replace white beeswax, making due allowance for color, but paraffin wax will not replace beeswax, even though its light color seems to place it above yellow beeswax.

And this leads to the third point: the use of good-quality ingredients, and ingredients of the correct quality. Ordinary lanolin is not the same thing as anhydrous lanolin. The replacement of one with the other, weight for weight, will give discouragingly different results. Use exactly what the formula calls for: if you are not acquainted with the substance and you are in doubt as to just what is meant, discard the formula and use one you understand. Buy your chemicals from reliable sources. Many ingredients are obtainable in a number of different grades: if the formula does not designate the grade, it is understood that the best grade is to be used. Remember that a formula and the directions can tell you only part of the story. Some skill is often required to attain success. Practice with a small batch in such cases until you are sure of your technique. Many examples can be cited. If the formula calls for steeping quince seed for 30 minutes in cold water, steeping for 1 hour may yield a mucilage of too thin a consistency. The originator of the formula may have used a fresher grade of seed, or his conception of what "cold" water means may be different from yours. You should have a feeling for the right degree of mucilaginousness, and if steeping the seed for 30 minutes fails to produce it, steep them longer until you get the right kind of mucilage. If you do not know what the right kind is, you will have to experiment until you find out. This is the reason for the recommendation to make small experimental batches until successful results are obtained. Another case is the use of dyestuffs for coloring lotions and the like. Dyes vary in strength; they are all very powerful in tinting value; it is not always easy to state in quantitative terms how much to use. You must establish the quantity by carefully adding minute quantities until you have the desired tint. Gum tragacanth is one of those products which can give much trouble. It varies widely in solubility and bodying power; the quantity listed in the formula may be entirely unsuitable for your grade of tragacanth. Therefore, correction is necessary, which can be made only after experiments with the available gum.

In short, if you are completely inexperienced, you can profit greatly by experimenting. Such products as mouth washes, hair tonics, and astringent lotions need little or no experience, because they are, as a rule, merely mixtures of simple liquid and solid ingredients, which dissolve without difficulty and the end product is a clear solution that is ready for use when mixed. However, face creams, tooth pastes, lubricating greases, wax polishes, etc., whose formulation requires relatively elaborate procedure and which must have a definite final viscosity, need some skill and not infrequently some experience.

### *Figuring*

Some prefer proportions expressed by weight or volume, others use percentages. In different industries and foreign countries different systems of weights and measures are used. For this reason, no one set of units could be satisfactory for everyone. Thus divers formulae appear with different units, in accordance with their sources of origin. In some cases, parts are given instead of percentage or weight or volume. On the pages preceding the index, conversion tables of weights and measures are listed. These are used for changing from one system to another. The following examples illustrate typical units:

EXAMPLE No. 1

Ink for Marking Glass

| | | | |
|---|---|---|---|
| Glycerin | 40 | Ammonium Sulfate | 10 |
| Barium Sulfate | 15 | Oxalic Acid | 8 |
| Ammonium Bifluoride | 15 | Water | 12 |

Here no units are mentioned. In this case, it is standard practice to use parts by weight throughout. Thus here we may use ounces, grams, pounds, or kilograms as desired. But if ounces are used for one item, the ounce must be the unit for all the other items in the formula.

EXAMPLE NO. 2

Flexible Glue

| | | | |
|---|---|---|---|
| Powdered Glue | 30.90% | Glycerin | 5.15% |
| Sorbitol (85%) | 15.45% | Water | 48.50% |

Where no units of weight or volume, but percentages are given, forget the percentages and use the same method as given in Example No. 1.

EXAMPLE NO. 3

Antiseptic Ointment

| | | | |
|---|---|---|---|
| Petrolatum | 16 parts | Benzoic Acid | 1 part |
| Coconut Oil | 12 parts | Chlorothymol | 1 part |
| Salicylic Acid | 1 part | | |

The instructions given for Example No. 1 also apply to Example No. 3. In many cases, it is not wise to make up too large a quantity of a product before making a number of small batches to first master the necessary technique and also to see whether the product is suitable for the particular purpose for which it is intended. Since, in many cases, a formula may be given in proportions as made up on a factory scale, it is advisable to reduce the quantities proportionately.

EXAMPLE NO. 4

Neutral Cleansing Cream

| | | | |
|---|---|---|---|
| Mineral Oil | 80 lb | Water | 90 lb |
| Spermaceti | 30 lb | Glycerin | 10 lb |
| Glyceryl Monostearate | 24 lb | Perfume | To suit |

Here, instead of pounds, ounces or even grams may be used. This formula would then read:

| | | | |
|---|---|---|---|
| Mineral Oil | 80 g | Water | 90 g |
| Spermaceti | 30 g | Glycerin | 10 g |
| Glyceryl Monostearate | 24 g | Perfume | To suit |

Reduction in bulk may also be obtained by taking the same fractional part or portion of each ingredient in a formula. Thus in the following formula:

EXAMPLE No. 5

Vinegar Face Lotion

| | | | |
|---|---|---|---|
| Acetic Acid (80%) | 20 | Alcohol | 440 |
| Glycerin | 20 | Water | 500 |
| Perfume | 20 | | |

We can divide each amount by ten and then the finished bulk will be only one tenth of the original formula. Thus it becomes:

| | | | |
|---|---|---|---|
| Acetic Acid (80%) | 2 | Alcohol | 44 |
| Glycerin | 2 | Water | 50 |
| Perfume | 2 | | |

*Apparatus*

For most preparations, pots, pans, china, and glassware, which are used in every household, will be satisfactory. For making fine mixtures and emulsions, a malted- ıilk mixer or egg beater is necessary. For weighing, a small, low-priced scale should be purchased from a laboratory-supply house. For measuring fluids, glass graduates or measuring glasses may be purchased from your local druggist. Where a thermometer is necessary, a chemical thermometer should be obtained from a druggist or chemical-supply firm.

*Methods*

To understand better the products which you intend to make, it is advisable that you read the complete section covering such products. You may learn different methods that may be used and also to avoid errors which many beginners are prone to make.

*Containers for Compounding*

Where discoloration or contamination is to be avoided, as in light-colored, or food and drug products, it is best to use enameled or earthenware vessels. Aluminum is also highly desirable in such cases, but it should not be used with alkalis as these dissolve and corrode aluminum.

*Heating*

To avoid overheating, it is advisable to use a double boiler when temperatures below 212°F (temperature of boiling water) will suffice. If a double boiler is not at hand, any pot may be filled with water and the vessel containing the ingredients to be heated placed in the water. The pot may then be heated by any flame without fear of overheating. The water in the pot, however, should be replenished from time to time; it must not be allowed to "go dry." To get uniform higher temperatures, oil, grease, or wax is used in the outer container in place of water. Here, of course, care must be taken to stop heating when thick fumes are given off as these are inflammable. When higher uniform temperatures are necessary, molten lead may be used as a heating medium. Of course, with chemicals which melt uniformly and are nonexplosive, direct heating over an open flame is permissible, with stirring, if necessary.

Where instructions indicate working at a certain temperature, it is important to attain the proper temperature not by guesswork, but by the use of a thermometer. Deviations from indicated temperatures will usually result in spoiled preparations.

### *Temperature Measurement*

In the United States and in Great Britain, the Fahrenheit scale of temperature is used. The temperature of boiling water is 212° Fahrenheit (212°F); the temperature of melting ice is 32° Fahrenheit (32°F).

In scientific work, and in most foreign countries, the Centigrade scale is used, on which the temperature of boiling water is 100 °Centigrade (100°C) and the temperature of melting ice is 0° Centigrade (0°C).

The temperature of liquids is measured by a glass thermometer. This is inserted as deeply as possible in the liquid and is moved about until the temperature reading remains steady. It takes a short time for the glass of the thermometer to reach the temperature of the liquid. The thermometer should not be placed against the bottom or side of the container, but near the center of the liquid in the vessel. Since the glass of the thermometer bulb is very thin, it breaks easily when striking it against any hard surface. A cold thermometer should be warmed gradually (by holding it over the surface of a hot liquid) before immersion. Similarly the hot thermometer when taken out of the liquid should not be put into cold water suddenly. A sharp change in temperature will often crack the glass.

### *Mixing and Dissolving*

Ordinary dissolution (e.g., that of sugar in water) is hastened by stirring and warming. Where the ingredients are not corrosive, a clean stick, a fork, or spoon may be used as a stirring rod. These may also be used for mixing thick creams or pastes. In cases where very thorough stirring is necessary (e.g., in making mayonnaise, milky polishes, etc.), an egg beater or a malted-milk mixer is necessary.

### *Filtering and Clarification*

When dirt or undissolved particles are present in a liquid, they are removed by settling or filtering. In the first procedure, the solution is allowed to stand and if the particles are heavier than the liquid they will gradually sink to the bottom. The liquid may be poured or siphoned off carefully and, in some cases, it is then sufficiently clear for use. If, however, the particles do not settle out, then they must be filtered off. If the particles are coarse they may be filtered or strained through muslin or other cloth. If they are very small, filter paper is used. Filter papers may be obtained in various degrees of fineness. Coarse filter paper filters rapidly but will not retain extremely fine particles. For fine particles, a very fine grade of filter paper should be used. In extreme cases, even this paper may not be fine enough. Then, it will be necessary to add to the liquid 1 to 3% infusorial earth or magnesium carbonate. These are filter aids that clog up the pores of the filter paper and thus reduce their size and hold back undissolved material of extreme fineness. In all such filtering, it is advisable to take the first portions of the filtered liquid and pour them through the filter again as they may develop cloudiness on standing.

### *Decolorizing*

The most commonly used decolorizer is decolorizing carbon. This is added to the liquid to the extent of 1 to 5% and the liquid is heated, with stirring, for ½ hour to as high a temperature as is feasible. The mixture is then allowed to stand for a while and filtered. In some cases, bleaching must be resorted to.

### *Pulverizing and Grinding*

Large masses or lumps are first broken up by wrapping in a clean cloth, placing between two boards, and pounding with a hammer. The smaller pieces are then pounded again to reduce their size. Finer grinding is done in a mortar with a pestle.

### *Spoilage and Loss*

All containers should be closed when not in use to prevent evaporation or contamination by dust; also because, in some cases, air affects the product adversely. Many chemicals attack or corrode the metal containers in which they are kept. This is particularly true of liquids. Therefore, liquids should be transferred into glass bottles which should be as full as possible. Corks should be covered with aluminum foil (or dipped in melted paraffin wax when alkalis are present).

Glue, gums, olive oil, or other vegetable or animal products may ferment or become rancid. This produces discoloration or unpleasant odors. To avoid this, suitable antiseptics or preservatives must be used. Cleanliness is of utmost importance. All containers must be cleaned thoroughly before use to avoid various complications.

### *Weighing and Measuring*

Since, in most cases, small quantities are to be weighed, it is necessary to get a light scale. Heavy scales should not be used for weighing small amounts as they are not accurate enough for this type of weighing.

For measuring volumes of liquids, measuring glasses or cylinders (graduates) should be used. Since this glassware cracks when heated or cooled suddenly, it should not be subjected to sudden changes of temperature.

### *Caution*

Some chemicals are corrosive and poisonous. In many cases, they are labeled as such. As a precautionary measure, it is advised not to inhale them and, if smelling is absolutely necessary, only to sniff a few inches from the cork or stopper. Always work in a well-ventilated room when handling poisonous or unknown chemicals. If anything is spilled, it should be wiped off and washed away at once.

### *Where to Buy Chemicals and Apparatus*

Many chemicals and most glassware can be purchased from your druggist. A list of suppliers of all products is at the end of this book.

### *Advice*

This book is the result of cooperation of many chemists and engineers who have given freely of their time and knowledge. It is their business to act as consultants and to give advice on technical matters for a fee. As publishers, we do not maintain a laboratory or consulting service to compete with them.

Please, therefore, do not ask us for advice or opinions, but confer with a chemist in your vicinity.

### *Extra Reading*

Keep up with new developments of materials and methods by reading technical magazines. Many technical publications are listed under references in the back of this book.

### *Calculating Costs*

Raw materials, purchased in small quantities, are naturally higher in price than when bought in large quantities. Commercial prices, as given in the trade papers and catalogs of manufacturers, are for large quantities such as barrels, drums, or sacks. For example, 1 lb. epsom salts, bought at retail, may cost 10 or 15 cents. In barrel lots its price is about 2 or 3 cents per pound.

Typical Costing Calculation

Formula for Beer- or Milk-Pipe Cleaner

| | | | | | |
|---|---|---|---|---|---|
| Soda Ash | 25 lb | @ | $0.02½ | per lb = | $ 0.63 |
| Sodium Perborate | 75 lb | @ | 0.16 | per lb = | 12.00 |
| Total | 100 lb | | | Total | $12.63 |

If 100 lb cost $12.63, 1 lb will cost $12.63 divided by 100 or about $0.126, assuming no loss.

Always weigh the amount of finished product and use this weight in calculating costs. Most compounding results in some loss of material because of spillage, sticking to apparatus, evaporation, etc. Costs of making experimental lots are always high and should not be used for figuring costs. To meet competition, it is necessary to buy in large quantities and manufacturing costs should be based on these.

## Elementary Preparations

The simple recipes that follow have been selected because of their importance and because they are easy to make.

The succeeding chapters go into greater detail and give many different types and modifications of these and other recipes for home and commercial use.

### *Cleansing Creams*

Cleansing creams, as the name implies, serve as skin cleaners. Their basic ingredients are oils and waxes which are rubbed into the skin. When wiped off, they carry off dirt and dead skin. The liquefying type cleansing cream contains no matter and melts or liquefies when rubbed on the skin. To suit different climates and likes and dislikes harder or softer products can be made.

Cleansing Cream (Liquefying)

| | |
|---|---|
| Liquid Petrolatum | 5.5 |
| Paraffin Wax | 2.5 |
| Petrolatum | 2.0 |

Melt the ingredients together, with stirrings, in an aluminum or enamelled dish and allow to cool. Then stir in a perfume oil. Allow to stand until it becomes hazy and then pour into jars, which should be allowed to stand undisturbed overnight.

### *Cold Creams*

The most important facial cream is the cold cream. This type of cream contains mineral oil and wax which are emulsified in water with a small amount of borax or glycosterin. The function of a cold cream is to form a film that takes up dirt and waste tissue, which are removed when the skin is wiped thoroughly. Many modifications of this basic cream are encountered in stores. They vary in color, odor, and in claims, but, essentially, they are not more useful than this simple cream. The latest type of cold cream is the nongreasy cold cream which is of particular interest because it is nonalkaline and, therefore, nonirritating for sensitive skins.

Cold Cream

| | |
|---|---|
| Liquid Petrolatum | 52 g |
| White Beeswax | 14 g |

Heat this in an aluminum or enamelled double boiler. (The water in the outer pot should be brought to a boil.) In a separate aluminum or enamelled pot dissolve:

| | |
|---|---|
| Borax | 1 g |
| Water | 33 cc |

and bring this to a boil. Add this in a thin stream to the melted wax, while stirring vigorously in one direction only. When the temperature drops to 140°F, add 0.5 cc perfume oil and continue stirring until the temperature drops to 120°F. At this point, pour into jars, where the cream will set after a while. If a harder cream is desired, reduce the amount of liquid petrolatum. If a softer cream is wanted, increase it.

Nongreasy Cold Cream

| | |
|---|---|
| White Paraffin Wax | 1.25 |
| Petrolatum | 1.50 |
| Glycosterin or Glyceryl Monostearate | 2.25 |
| Liquid Petrolatum | 3.00 |

Heat this mixture in an aluminum or enamelled double boiler. (The water in the outer pot should be boiling.) Stir until clear. To this slowly add, while stirring vigorously:

| | |
|---|---|
| Boiling Water | 10 |

Continue stirring until smooth and then add, with stirring, perfume oil. Pour into jars at 110 to 130°F and cover the jars as soon as possible.

*Vanishing Creams*

Vanishing creams are nongreasy soapy creams which have a cleansing effect. They are also used as a powder base.

Vanishing Cream

| | |
|---|---|
| Stearic Acid | 18 oz |

Melt this in an aluminum or enamelled double boiler. (The water in the outer pot must be boiling.) Add, in a thin stream, while stirring vigorously, the following boiling solution made in an aluminum or enamelled pot:

| | |
|---|---|
| Potassium Carbonate | ¼ oz |
| Glycerin | 6½ oz |
| Water | 5 lb |

Continue stirring until the temperature falls to 135°F, then mix in a perfume oil and stir from time to time until cold. Allow to stand overnight and stir again the next day. Pack into jars and close these tightly.

*Hand Lotions*

Hand lotions are usually clear or milky liquids or salves which are useful in protecting the skin from roughness and redness because of exposure to cold, hot water, soap, and other agents. Chapped hands are common. The use of a good hand lotion keeps the skin smooth, soft, and in a healthy condition. The lotion is best applied at night, rather freely, and cotton gloves may be worn to prevent soiling. During the day, it should be put on sparingly and the excess wiped off.

Hand Lotion
(Salve)

| | |
|---|---|
| Boric Acid | 1 |
| Glycerin | 6 |

Warm these in an aluminum or enamelled dish and stir until dissolved (clear). Then allow to cool and work this liquid into the following mixture, adding only a little at a time.

| | |
|---|---|
| Lanolin | 6 |
| Petrolatum | 8 |

To impart a pleasant odor a little perfume may be added and worked in.

Hand Lotion
(Milky liquid)

| | | |
|---|---|---|
| Lanolin | ¼ | tsp |
| Glycosterin or Glyceryl Monostearate | 1 | oz |
| Tincture of Benzoin | 2 | oz |
| Witch Hazel | 25 | oz |

Melt the first two items together in an aluminum or enamelled double boiler. If no double boiler is at hand, improvise one by placing a dish in a small pot containing boiling water. When the mixture becomes clear, remove from the double boiler and add slowly, while stirring vigorously, the tincture of benzoin and then the witch hazel. Continue stirring until cool and then put into one or two large bottles and shake vigorously. The finished lotion is a milky liquid comparable to the best hand lotions on the market sold at high prices.

---

*Brushless Shaving Creams*

Brushless or latherless shaving creams are soapy in nature and do not require lathering or water. The formula given here is of the latest type being free from alkali and nonirritating. It should be borne in mind, however, that certain beards are not softened by this type of cream and require the old-fashioned lathering shaving cream.

Brushless Shaving Cream

| | |
|---|---|
| White Mineral Oil | 10 |
| Glycosterin or Glyceryl Monostearate | 10 |
| Water | 50 |

Heat the first two ingredients together in a Pyrex or enamelled dish to 150°F and run in slowly, while stirring, the water which has been heated to boiling. Allow to cool to 150°F and, while stirring, add a few drops of perfume oil. Continue stirring until cold.

---

*Mouth Washes*

Mouth washes and oral antiseptics are of practically negligible value. However, they are used because of their refreshing taste and slight deodorizing effect.

Mouth Wash

| | |
|---|---|
| Benzoic Acid | 5/8 |
| Tincture of Rhatany | 3 |
| Alcohol | 20 |
| Peppermint Oil | 1/8 |

Mix together in a dry bottle until the benzoic acid is dissolved. One teaspoonful is used to a small-wine-glassful of water.

---

*Tooth Powders*

The cleansing action of tooth powders depends on their contents of soap and mild abrasives, such as precipitated chalk and magnesium carbonate. The antiseptic present is practically of no value. The flavoring ingredients mask the taste of the soap and give the mouth a pleasant aftertaste.

Tooth Powder

| | |
|---|---|
| Magnesium Carbonate | 420 g |
| Precipitated Chalk | 565 g |
| Sodium Perborate | 55 g |
| Sodium Bicarbonate | 45 g |
| Powdered White Soap | 50 g |
| Powdered Sugar | 90 g |
| Wintergreen Oil | 8 cc |
| Cinnamon Oil | 2 cc |
| Menthol | 1 g |

Dissolve the last three ingredients together and then rub well into the sugar. Add the soap and perborate, mixing well. Add the chalk, with good mixing, and then the sodium bicarbonate and magnesium carbonate. Mix thoroughly and sift through a fine wire screen. Keep dry.

---

*Foot Powders*

Foot powders consist of talc or starch with or without an antiseptic or deodorizer. In the following formula the perborates liberate oxygen, when in contact with perspiration, which tends to destroy unpleasant odors. The talc acts as a lubricant and prevents friction and chafing.

Foot Powder

| | |
|---|---|
| Sodium Perborate | 3 |
| Zinc Peroxide | 2 |
| Talc | 15 |

Mix thoroughly in a dry container until uniform. This powder must be kept dry or it will spoil.

---

*Liniments*

Liniments usually consist of an oil and an irritant, such as methyl salicylate or turpentine. The oil acts as a solvent and tempering agent for the irritant. The irritant produces a rush of blood and warmth which is often slightly helpful.

Sore-Muscle Liniment

| | |
|---|---|
| Olive Oil | 6 fl oz |
| Methyl Salicylate | 3 fl oz |

Mix together and keep in a well-stoppered bottle. Apply externally, but do not use on chafed or cut skin.

---

*Chest Rubs*

In spite of the fact that chest rubs are practically useless, countless sufferers use them. Their action is similar to that of liniments and they differ only in that they are in the form of a salve.

Chest-Rub Salve

| | |
|---|---|
| Yellow Petrolatum | 1 lb |
| Paraffin Wax | 1 oz |
| Eucalyptus Oil | 2 fl oz |
| Menthol | ½ oz |
| Cassia Oil | ⅛ fl oz |
| Turpentine | ½ fl oz |

Melt the petrolatum and paraffin wax together in a double boiler and then add the menthol. Remove from the heat, stir, and cool a little; then mix in the oils, and turpentine. When it begins to thicken, pour into tins and cover.

---

*Insect Repellents*

Preparations of this type may irritate sensitive skins and they will not always work.

Mosquito-Repelling Oil

| | |
|---|---|
| Cedar Oil | 2 fl oz |
| Citronella Oil | 4 fl oz |
| Spirits of Camphor | 8 fl oz |

Mix in a dry bottle and the oil is ready for use. This preparation may be smeared on the skin as often as is necessary.

---

*Fly Sprays*

Fly sprays usually consist of deodorized kerosene, perfume, and an active insecticide. In some cases, they merely stun the flies who may later recover and begin buzzing again.

Fly Spray

| | |
|---|---|
| Deodorized Kerosene | 80 fl oz |
| Methyl Salicylate | 1 fl oz |
| Pyrethrum Powder | 10 oz |

Mix thoroughly by stirring from time to time; allow to stand covered overnight and then filter through muslin.

This spray is inflammable and should not be used near open flames.

---

Deodorant Spray

(For public buildings, sick rooms, lavatories, etc.)

| | |
|---|---|
| Pine-Needle Oil | 2 |
| Formaldehyde | 2 |
| Acetone | 6 |

* Isopropyl Alcohol 20

One ounce of this mixture is diluted with 1 pt. water for spraying.

---

Cresol Disinfectant

† Caustic Soda 25.5 g
Water 140.0 cc

Dissolve in a Pyrex or enamelled dish and warm. To this, add slowly the following warmed mixture:

‡ Cresylic Acid 500.0 cc
Rosin 170.0 g

Stir until dissolved and add water to make 1,000 cc.

---

Ant Poison

Sugar 1 lb
Water 1 qt
‡ Arsenate of Soda 125 g

Boil and stir until uniform; strain through muslin and add 1 spoonful honey.

---

Bedbug Exterminator

* Kerosene 90 fl oz
Clove Oil 5 fl oz
‡ Cresol 1 fl oz
Pine Oil 4 fl oz

Simply mix and bottle.

---

Nonstaining Mothproofing Fluid

Sodium Aluminum Silicofluoride 0.50
Water 98.00
Glycerin 0.50
"Sulfatate" (Wetting Agent) 0.25

Stir until dissolved.

---

Fly Paper

Rosin 32
Rosin Oil 20
Castor Oil 8

Heat this mixture in an aluminum or enamelled pot on a gas stove, with stirring, until all the rosin has melted and dissolved. While hot, pour on firm paper sheets of suitable size which have been brushed with soap water just before coating. Smooth out the coating with a long knife or piece of thin flat wood and allow to cool. If a heavier coating is desired, increase the amount of rosin. Similarly, a thinner coating results by reducing the amount of rosin. The finished paper should be laid flat and not exposed to undue heat.

---

Baking Powder

Bicarbonate of Soda 28
Monocalcium Phosphate 35
Corn Starch 27

Mix these powders thoroughly

* Inflammable.
† Do not get this on the skin as it is corrosive.
‡ Poison.

in a dry can by shaking and rolling for ½ hour. Pack into dry airtight tins as moisture will cause lumping.

---

Malted-Milk Powder

| | |
|---|---|
| Powdered Malt Extract | 5 |
| Powdered Skim Milk | 2 |
| Powdered Sugar | 3 |

Mix thoroughly by shaking and rolling in a dry can. Pack in an airtight container.

---

Cocoa-Malt Powder

| | |
|---|---|
| Corn Sugar | 55 |
| Powdered Malt | 19 |
| Powdered Skim Milk | 12½ |
| Cocoa | 13 |
| Vanillin | ⅛ |
| Powdered Salt | ⅜ |

Mix thoroughly and then run through a fine wire sieve.

---

Sweet Cocoa Powder

| | |
|---|---|
| Cocoa | 17½ oz |
| Powdered Sugar | 32½ oz |
| Vanillin | ¾ g |

Mix thoroughly and sift.

---

Pure Lemon Extract

| | |
|---|---|
| Lemon Oil U.S.P. | 6½ fl oz |
| Alcohol | 121½ fl oz |

Shake together in 1-gal jug until dissolved.

Artificial Vanilla Flavor

| | |
|---|---|
| Vanillin | ¾ oz |
| Coumarin | ¼ oz |
| Alcohol | 2 pt |

Stir the ingredients in a glass or china pitcher until dissolved. Then mix into the following solution:

| | |
|---|---|
| Sugar | 12 oz |
| Water | 5¼ pt |
| Glycerin | 1 pt |

Color brown by adding sufficient burnt-sugar coloring.

---

Canary Food

| | |
|---|---|
| Dried and Chopped Egg Yolk | 2 |
| Poppy Heads (Coarse Powder) | 1 |
| Cuttlefish Bone (Coarse Powder) | 1 |
| Granulated Sugar | 2 |
| Powdered Soda Crackers | 8 |

Mix well together.

---

Blue-Black Writing Ink

| | |
|---|---|
| Naphthol Blue Black | 1 oz |
| Powdered Gum Arabic | ½ oz |
| Carbolic Acid | ¼ oz |
| Water | 1 gal |

Stir together in a glass or enamelled vessel until dissolved.

Indelible Laundry-Marking Ink

*a*

| | |
|---|---|
| Soda Ash | 1 oz |
| Powdered Gum Arabic | 1 oz |
| Water | 10 fl oz |

Stir until dissolved.

*b*

| | |
|---|---|
| Silver Nitrate | 4 oz |
| Powdered Gum Arabic | 4 oz |
| Lampblack | 2 oz |
| Water | 40 fl oz |

Stir this in a glass or porcelain dish until dissolved. Do not expose the mixture to strong light or it will spoil. Then pour into a brown glass bottle. In using these solutions, wet the cloth with solution *a* and allow to dry. Then write on it with solution *b* using a quill pen.

---

Green Marking Crayon

| | |
|---|---|
| Ceresin | 8 |
| Carnauba Wax | 7 |
| Paraffin Wax | 4 |
| Beeswax | 1 |
| Talc | 10 |
| Chrome Green | 3 |

Melt the first four ingredients in a container and then add the last two slowly, while stirring. Remove from the heat and continue stirring until thickening begins. Then pour into molds. If other-color crayons are desired, other pigments may be used. For example, for black, use carbon black or bone black; for blue, Prussian blue; for red, orange chrome yellow.

---

Antique Coloring for Copper

| | |
|---|---|
| Copper Nitrate | 4 oz |
| Acetic Acid | 1 oz |
| Water | 2 oz |

Dissolve by stirring together in a glass or porcelain vessel. Pack into glass bottles.

Wet the copper to be colored and apply the coloring solution hot.

---

Blue-Black Finish on Steel

*a* Place the object in molten sodium nitrate at 700 to 800°F for 2 to 3 minutes. Remove and allow to cool somewhat, wash in hot water, dry, and oil with mineral or linseed oil.

*b* Then put the object in the following solution for 15 minutes:

| | |
|---|---|
| Copper Sulfate | ½ oz |
| Iron Chloride | 1 lb |
| Hydrochloric Acid | 4 oz |
| Nitric Acid | ½ oz |
| Water | 1 gal |

Allow to dry for several hours. Place in a solution again for 15 minutes, remove and dry for 10 hours. Place in boiling water for ½ hour, dry, and scratch-brush very lightly. Oil with mineral or linseed oil and wipe dry.

Rust-Prevention Compound

| | |
|---|---|
| Lanolin | 1 |
| * Naphtha | 2 |

Mix until dissolved.

The metal to be protected is cleaned with a dry cloth and then coated with the composition.

---

Metal Polish

| | | |
|---|---|---|
| Naphtha | 62 | oz |
| Oleic Acid | ⅓ | oz |
| Abrasive | 7 | oz |
| Triethanolamine Oleate | ⅓ | oz |
| Ammonia (26°) | 1 | oz |
| Water | 1 | gal |

In one container mix together the naphtha and oleic acid to a clear solution. Dissolve the triethanolamine oleate in the water separately, stir in the abrasive, and then add the naphtha solution. Stir the resulting mixture at a high speed until a uniform creamy emulsion results. Then add the ammonia and mix well, but do not agitate so vigorously as before.

---

Glass-Etching Fluid

| | |
|---|---|
| Hot Water | 12 |
| † Ammonium Bifluoride | 15 |
| Oxalic Acid | 8 |
| Ammonium Sulfate | 10 |
| Glycerin | 40 |
| Barium Sulfate | 15 |

Warm the washed glass slightly before writing on it with this fluid. Allow the fluid to act on the glass for about 2 minutes.

---

Leather Preservative

| | |
|---|---|
| Cold-Pressed Neatsfoot Oil | 10 |
| Castor Oil | 10 |

Mix.

This is an excellent preservative for leather book bindings, luggage, and other leather goods.

---

White-Shoe Dressing

| | | |
|---|---|---|
| Lithopone | 19 | oz |
| Titanium Dioxide | 1 | oz |
| Bleached Shellac | 3 | oz |
| Ammonium Hydroxide | ¼ | fl oz |
| Water | 25 | fl oz |
| Alcohol | 25 | fl oz |
| Glycerin | 1 | oz |

Dissolve the last four ingredients by mixing in a porcelain vessel. When dissolved, stir in the first two pigments. Keep in stoppered bottles and shake before using.

---

Waterproofing for Shoes

| | |
|---|---|
| Wool Grease | 8 |
| Dark Petrolatum | 4 |
| Paraffin Wax | 4 |

Melt together in any container.

* Inflammable — keep away from flames.

† Corrosive.

### *Polishes*

Polishes are generally used to restore the original luster and finish of a smooth surface. They are also expected to clean the surface and to prevent corrosion or deterioration. There is no one polish which will give good results on all surfaces.

Most polishes contain oil or wax for their lustering or polishing properties. Oil polishes are easy to apply, but the surfaces on which they are used attract dust and show finger marks. Wax polishes are more difficult to apply, but are more lasting.

Oil or wax polishes are of two types: waterless and aqueous. The former are clear or translucent, the latter are milky in appearance.

For use on metals, abrasives of various kinds, such as tripoli, silica dust, or infusorial earth, are incorporated to grind away oxide films or corrosion products.

Black Shoe Polish

| | |
|---|---|
| Carnauba Wax | 5½ oz |
| Crude Montan Wax | 5½ oz |

Melt together in a double boiler. (The water in the outer container should be boiling.) Then stir in the following melted and dissolved mixture:

| | |
|---|---|
| Stearic Acid | 2 oz |
| Nigrosine Base | 1 oz |

Then stir in

| | |
|---|---|
| Ceresin | 15 oz |

Remove all flames and run in slowly, while stirring,

| | |
|---|---|
| Turpentine | 90 fl oz |

Allow the mixture to cool to 105°F. and pour into airtight tins which should stand undisturbed overnight.

Clear Oil-Type Auto Polish

| | |
|---|---|
| Paraffin Oil | 5 pt |
| Raw Linseed Oil | 2 pt |
| China-Wood Oil | ½ pt |
| * Benzol | ¼ pt |
| * Kerosene | ¼ pt |
| Amyl Acetate | 1 tbsp |

Mix together in a glass jar and keep it stoppered.

Paste-Type Auto and Floor Wax

| | |
|---|---|
| Yellow Beeswax | 1 oz |
| Ceresin | 2½ oz |
| Carnauba Wax | 4½ oz |
| Montan Wax | 1¼ oz |
| * Naphtha or Mineral Spirits | 1 pt |
| * Turpentine | 2 oz |
| Pine Oil | ½ oz |

Melt the waxes together in a double boiler. Turn off the heat and run in the last three ingredients in a thin stream, with stirring. Pour into cans, cover, and allow to stand undisturbed overnight.

* Inflammable — keep away from flames.

Oil-and-Wax Type Furniture Polish

| | |
|---|---|
| Paraffin Oil | 1 pt |
| Powdered Carnauba Wax | ¼ oz |
| Ceresin Wax | ⅛ oz |

Heat together until all of the wax is melted. Allow to cool and pour into bottles before the mixture turns cloudy.

Liquid Polishing Wax

| | |
|---|---|
| Yellow Beeswax | 1 oz |
| Ceresin Wax | 4 oz |

Melt together and then cool to 130°F.; turn off all flames and stir in slowly:

| | |
|---|---|
| * Turpentine | 17 fl oz |
| Pine Oil | ½ fl oz |

Pour into cans or bottles which are closed tightly to prevent evaporation.

Floor Oil

| | |
|---|---|
| Mineral Oil | 46 fl oz |
| Beeswax | ½ oz |
| Carnauba Wax | 1 oz |

Heat together in double boiler until dissolved (clear). Turn off the flame and stir in

| | |
|---|---|
| * Turpentine | 3 fl oz |

---

*Lubricants*

Lubricants, in the form of oils or greases, are used to prevent friction and wearing of parts which are rubbed together. Lubricants must be chosen to fit specific uses. They consist of oils and fats often compounded with soaps and other unctuous substances. For heavy duty, heavy oils or greases are used and light oils are suitable for light duty.

Gum Lubricant

| | |
|---|---|
| White Petrolatum | 15 oz |
| Acid-Free Bone Oil | 5 oz |

Warm gently and mix together.

Graphite Grease

| | |
|---|---|
| Ceresin | 7 oz |
| Tallow | 7 oz |

Warm together and gradually work in with a stick:

| | |
|---|---|
| Graphite | 3 oz |

Stir until uniform and pack in tins when thickening begins.

Penetrating Oil
(For loosening rusted bolts, screws, etc.)

| | |
|---|---|
| Kerosene | 2 oz |
| Thin Mineral Oil | 7 oz |
| Secondary Butyl Alcohol | 1 oz |

Mix and keep in a stoppered bottle.

---

Molding Compound

| | |
|---|---|
| White Glue | 13 lb |
| Rosin | 13 lb |
| Raw Linseed Oil | ⅓ qt |
| Glycerin | 1 qt |
| Whiting | 19 lb |

Heat the white glue until it

* Inflammable.

melts. Then cook separately the rosin and raw linseed oil until the rosin is dissolved. Add the rosin, oil, and glycerin to the glue, stirring in the whiting until the mass reaches the consistency of a putty. Keep the mixture hot.

Press this mass into the die firmly and allow it to cool slightly before removing. The finished product is ready to use within a few hours after removal. Suitable pigments may be added to secure brown, red, black, or other color.

In applying ornaments made of this composition to a wood surface, they are first steamed to make them flexible; in this condition, they will adhere to the wood easily and securely. They can be bent to any shape, and no nails are required for applying them.

Grafting Wax

| | |
|---|---|
| Wool Grease | 11 |
| Rosin | 22 |
| Paraffin Wax | 6 |
| Beeswax | 4 |
| Japan Wax | 1 |
| Rosin Oil | 9 |
| Pine Oil | 1 |

Melt together until clear and pour into tins. This composition can be made thinner by increasing the amount of rosin oil and thicker by decreasing it.

Candles

| | |
|---|---|
| Paraffin Wax | 30.0 |
| Stearic Acid | 17.5 |
| Beeswax | 2.5 |

Melt together and stir until clear. If colored candles are desired, add a very small amount of any oil-soluble dye. Pour into vertical molds in which wicks are hung.

---

*Adhesives*

Adhesives are sticky substances used to unite two surfaces. Adhesives are specifically called glues, pastes, cements, mucilages, lutes, etc. For different uses different types are required.

Wall-Patching Plaster

| | |
|---|---|
| Plaster of Paris | 32 |
| Dextrin | 4 |
| Pumice Powder | 4 |

Mix thoroughly by shaking and rolling in a dry container. Keep away from moisture.

Cement-Floor Hardener

| | |
|---|---|
| Magnesium Fluosilicate | 1 lb |
| Water | 15 pt |

Mix until dissolved.

The cement should first be washed with clean water and then drenched with this solution.

Paperhanger's Paste

| | |
|---|---|
| White or Fish Glue | 4 oz |
| Cold Water | 8 oz |

| | |
|---|---|
| Venice Turpentine | 2 fl oz |
| Rye Flour | 1 lb |
| Cold Water | 16 fl oz |
| Boiling Water | 64 fl oz |

Soak the glue in the first amount of cold water for 4 hours. Dissolve on a water-bath (glue-pot) and while hot stir in the Venice turpentine. Use a cheap grade of rye or wheat flour, mix thoroughly with the second amount of cold water to about the consistency of dough or a little thinner, being careful to remove all lumps. Stir in 1 tbsp of powdered alum to 1 qt flour, then pour in the boiling water, stirring rapidly until the flour is thoroughly cooked. Let this cool and finally add the glue solution. This makes a very strong paste which will also adhere to a painted surface, owing to the Venice turpentine content.

Aquarium Cement

| | |
|---|---|
| Litharge | 10 |
| Plaster of Paris | 10 |
| Powdered Rosin | 1 |
| Dry White Sand | 10 |
| Boiled Linseed Oil | Sufficient |

Mix all together in the dry state, and make a stiff putty with the oil just before use.

Do not fill the aquarium for 3 days after cementing. This cement hardens under water, and will stick to wood, stone, metal, or glass and as it resists the action of sea water, it is useful for marine *aquaria*.

Wood-Dough Plastic

| | |
|---|---|
| * Collodion | 86 |
| Powdered Ester Gum | 9 |
| Wood Flour | 30 |

Allow the first two ingredients to stand until dissolved, stirring from time to time. Then, while stirring, add the wood flour, a little at a time, until uniform. This product can be made softer by adding more collodion.

Putty

| | |
|---|---|
| Whiting | 80 |
| Raw Linseed Oil | 16 |

Rub together until smooth. Keep in a closed container.

---

Wood-Floor Bleach

| | |
|---|---|
| Sodium Metasilicate | 90 |
| Sodium Perborate | 10 |

Mix thoroughly and keep dry in a closed can. Use 1 lb to 1 gal boiling water. Mop or brush on the floor, allow to stand ½ hour, then rub off and rinse well with water.

---

* Paint Remover

| | |
|---|---|
| Benzol | 5 pt |
| Ethyl Acetate | 3 pt |
| Butyl Acetate | 2 pt |
| Paraffin Wax | ½ lb |

* Inflammable.

Stir together until dissolved.

---

*Soaps and Cleaners*

Soaps are made from a fat or fatty acid and an alkali. They lather and produce a foam which entraps dirt and grease. There are many kinds of soaps.

Cleaners contain a solvent, such as naphtha, with or without a soap. Abrasive cleaners are soap pastes containing powdered pumice, stone, silica, etc.

Concentrated Liquid Soap

| | |
|---|---|
| Water | 11 |
| * Solid Caustic Potash | 1 |
| Glycerin | 4 |
| Red Oil (Oleic Acid) | 4 |

Dissolve the caustic soda in water, add the glycerin, and bring to a boil in an enamelled pot. Remove from the heat, add the red oil slowly, while stirring. If a more neutral soap is wanted, use more red oil.

Saddle Soap

| | |
|---|---|
| Beeswax | 5.0 |
| * Caustic Potash | 0.8 |
| Water | 8.0 |

Boil for 5 minutes, while stirring. In another vessel heat:

| | |
|---|---|
| Castile Soap | 1.6 |
| Water | 8.0 |

Mix the two solutions with good stirring; remove from the heat and add, while stirring:

| | |
|---|---|
| Turpentine | 12 |

Mechanics' Hand-Soap Paste

| | |
|---|---|
| Water | 1.8 qt |
| White Soap Chips | 1.5 lb |
| Glycerin | 2.4 oz |
| Borax | 6.0 oz |
| Dry Sodium Carbonate | 3.0 oz |
| Coarse Pumice Powder | 2.2 lb |
| Safrol | To suit |

Dissolve the soap in two thirds of the water by heat. Dissolve the last three ingredients in the rest of the water. Pour the two solutions together and stir well. When it begins to thicken, sift in the pumice, stirring constantly till thick, then pour into cans. Vary the amount of water, for heavier or softer paste. Water cannot be added to the finished soap.

Dry-Cleaning Fluid

| | |
|---|---|
| Glycol Oleate | 2 fl oz |
| Carbon Tetrachloride | 60 fl oz |
| Naphtha | 20 fl oz |
| Benzine | 18 fl oz |

This is an excellent cleaner that will not injure the finest fabrics.

Wall-Paper Cleaner

| | |
|---|---|
| Whiting | 10 lb |
| Calcined Magnesia | 2 lb |

* Do not get on the skin as it is corrosive.

| | |
|---|---|
| Fuller's Earth | 2 lb |
| Powdered Pumice | 12 oz |
| Lemenone or Citronella Oil | 4 oz |

Mix well together.

Household Cleaner

| | |
|---|---|
| Soap Powder | 2 |
| Soda Ash | 3 |
| Trisodium Phosphate | 40 |
| Finely Ground Silica | 55 |

Mix well and pack in the usual containers.

Window Cleanser

| | |
|---|---|
| Castile Soap | 2 |
| Water | 5 |
| Chalk | 4 |
| French Chalk | 3 |
| Tripoli Powder | 2 |
| Petroleum Spirits | 5 |

Mix well and pack in tight containers.

Straw-Hat Cleaner

Sponge the hat with a solution of:

| | |
|---|---|
| Sodium Hyposulfite | 10 oz |
| Glycerin | 5 oz |
| Alcohol | 10 oz |
| Water | 75 oz |

Lay the hat aside in a damp place for 24 hours and then apply a mixture of:

| | |
|---|---|
| Citric Acid | 2 oz |
| Alcohol | 10 oz |
| Water | 90 oz |

Press with a moderately hot iron after stiffening with gum water, if necessary.

Grease, Oil, Paint, and Lacquer Spot Remover

| | |
|---|---|
| Alcohol | 1 |
| Ethyl Acetate | 2 |
| Butyl Acetate | 2 |
| Toluol | 2 |
| Carbon Tetrachloride | 3 |

Place the garment with the spot over a piece of clean paper or cloth and wet the spot with this fluid; rub with a clean cloth toward the center of the spot. Use a clean section of cloth for rubbing and clean paper or cloth for each application of the fluid. This cleaner is inflammable and should be kept away from flames. Cleaners of this type should be used out of doors or in well-ventilated rooms as the fumes are toxic.

Paint-Brush Cleaner

| | | |
|---|---|---|
| *a* | Kerosene | 2.00 |
| | Oleic Acid | 1.00 |
| *b* | Strong Liquid Ammonia (28%) | 0.25 |
| | Denatured Alcohol | 0.25 |

Slowly stir *b* into *a* until a smooth mixture results. To clean brushes, pour into a can and leave the brushes in it overnight. In the morning, wash out with warm water.

Rust and Ink Remover

Immerse the part of the fabric with the rust or ink spot alternately in solutions *a* and *b*, rinsing with water after each immersion.

| | | |
|---|---|---|
| *a* | Ammonium Sulfide Solution | 1 |
| | Water | 19 |
| *b* | * Oxalic Acid | 1 |
| | Water | 19 |

Javelle Water
(Laundry bleach)

| | |
|---|---|
| Bleaching Powder | 2 oz |
| Soda Ash | 2 oz. |
| Water | 5 gal. |

Mix well until the reaction is completed. Allow to settle overnight and siphon off the clear liquid.

Liquid Laundry Blue

| | |
|---|---|
| Prussian Blue | 1 |
| Distilled Water | 32 |
| * Oxalic Acid | ¼ |

Dissolve by mixing in a crock or wooden tub.

Glassine Paper

Paper is coated with or dipped in the following solution and then hung up to dry.

| | |
|---|---|
| Copal Gum | 10 oz |
| Alcohol | 30 fl oz |
| Castor Oil | 1 fl oz |

Dissolve by letting stand overnight in a covered jar and stirring the next day.

* Poisonous.

Waterproofing Paper and Fiberboard

The following composition and method of application will make uncalendered paper, fiberboard, and similar porous material waterproof.

| | |
|---|---|
| Paraffin (M.P. about 130°F.) | 22.5 |
| Trihydroxyethylamine Stearate | 3.0 |
| Water | 74.5 |

The paraffin wax is melted and the stearate added to it. The water is then heated to nearly the boiling point and vigorously agitated with a suitable mechanical stirring device while the mixture of melted wax and emulsifier is being slowly added. This mixture is cooled while it is stirred.

The paper or fiberboard is coated on the side which is to be in contact with water. This method works most effectively on paper-pulp molded containers and has the advantage of being much cheaper than dipping in melted paraffin as only about one tenth as much paraffin is needed. In addition, the outside of the container is not greasy and can be printed on after treatment which is not the case when treating with melted wax.

### * Waterproofing Liquid

| | |
|---|---|
| Paraffin Wax | 2/5 oz |
| Gum Dammar | 1 1/5 oz |
| Pure Rubber | /8 oz |
| Benzol | 13 oz |
| Carbon Tetrachloride | To make 1 gal |

Dissolve the rubber in the benzol, add the other ingredients, and allow to dissolve.

This liquid is suitable for wearing apparel and wood. It is applied by brushing on two or more coats, allowing each to dry before applying another coat. Apply outdoors as vapors are inflammable and toxic.

### Waterproofing Heavy Canvas

| | |
|---|---|
| Raw Linseed Oil | 1 gal |
| Crude Beeswax | 13 oz |
| White Lead | 1 lb |
| Rosin | 12 oz |

Heat, while stirring, until all lumps are removed and apply warm to the upper side of the canvas, wetting it with a sponge on the underside before application.

### Waterproofing Cement

| | |
|---|---|
| China-Wood Oil Fatty Acids | 10 oz |
| Paraffin Wax | 10 oz |
| Kerosene | 2½ gal |

Stir until dissolved. Paint or spray on cement walls, which must be dry.

### Oil- and Greaseproofing Paper and Fiberboard

This solution, applied by brush, spray, or dipping, will leave a thin film which is impervious to oil and grease. Applied to paper or fiber containers, it will enable them to retain oils and greases.

| | |
|---|---|
| Starch | 6.6 |
| Caustic Soda | 0.1 |
| Glycerin | 2.0 |
| Sugar | 0.6 |
| Water | 90.5 |
| Sodium Salicylate | 0.2 |

The caustic soda is dissolved in the water. Then the starch is made into a thick paste by adding a portion of this solution. The paste is then added to the water. The resulting mixture is placed on a water bath and heated to about 85°C, until all the starch granules have broken. The temperature is maintained about ½ hour longer at 85°C. The other substances are then added and thoroughly mixed. The composition is now ready for application. Less water may be used if applied hot and then a thicker coating will result.

* Inflammable.

### Fireproof Paper

| | |
|---|---|
| Ammonium Sulfate | 8.00 |
| Boric Acid | 3.00 |
| Borax | 1.75 |
| Water | 100.00 |

The ingredients are mixed together in a gallon jug by shaking until dissolved.

The paper to be treated is dipped into this solution in a pan, until uniformly saturated. It is then taken out and hung up to dry. Wrinkles can be prevented by drying between cloths in a press.

---

### Fireproofing Canvas

| | | |
|---|---|---|
| Ammonium Phosphate | 1 | lb |
| Ammonium Chloride | 2 | lb |
| Water | ½ | gal |

Impregnate with the solution; squeeze out the excess, and dry. Washing or exposure to rain will remove fireproofing salts.

---

### Fireproofing Light Fabrics

| | | |
|---|---|---|
| Borax | 10 | oz |
| Boric Acid | 8 | oz |
| Water | 1 | gal |

Impregnate, squeeze, and dry. Fabrics so impregnated must be treated again after washing or exposure to rain as the fireproofing salts wash out easily.

### Dry Fire Extinguisher

| | |
|---|---|
| Ammonium Sulfate | 15 |
| Sodium Bicarbonate | 9 |
| Ammonium Phosphate | 1 |
| Red Ochre | 2 |
| "Silex" | 23 |

Use powdered substances only. Mix well and pass through a fine sieve. Pack in tight containers to prevent lumping.

---

### Fire-Extinguishing Liquid

| | |
|---|---|
| Carbon Tetrachloride | 95 |
| Solvent Naphtha | 5 |

The naphtha minimizes the development of toxic fumes when extinguishing fires.

---

### Fire Kindler

| | |
|---|---|
| Rosin or Pitch | 10 |
| Sawdust | 10 or more |

Melt, mix, and cast in forms.

---

### Solidified Gasoline

| | | |
|---|---|---|
| * Gasoline | ½ | gal |
| Fine-Shaved White Soap | 12 | oz |
| Water | 1 | pt |
| Ammonia | 5 | oz |

Heat the water, add the soap, mix and, when cool, add the ammonia. Then slowly work in the gasoline to form a semisolid mass.

* Inflammable.

### Boiler Compound

| | |
|---|---|
| Soda Ash | 87 |
| Trisodium Phosphate | 10 |
| Starch | 1 |
| Tannic Acid | 2 |

Use powders, mix well, and then pass through a fine sieve.

---

### Noncorrosive Soldering Flux

| | |
|---|---|
| Powdered Rosin | 1 |
| Denatured Alcohol | 4 |

Soak overnight and mix well.

---

## *Photographic Solutions*

### Developing Solution

Stock Solution *a*

| | |
|---|---|
| Pyro | 4 oz |
| Pure Sodium Bisulfite | 280 gr |
| Potassium Bromide | 32 gr |
| Distilled Water | 64 oz |

Dissolve in a glass or enamel dish.

Stock Solution *b*

| | |
|---|---|
| Pure Sodium Sulfite | 7 oz |
| Pure Sodium Carbonate | 5 oz |
| Distilled Water | 64 oz |

Dissolve separately in a glass or enamel dish.

Use the following proportions:

| | |
|---|---|
| Stock Solution *a* | 2 |
| Stock Solution *b* | 2 |
| Distilled Water | 16 |

At 65°F., this developer requires about 8 minutes.

### Acid-Hardening Fixing Bath

| | | |
|---|---|---|
| *a* | Sodium Hyposulfite | 32 |
| | Distilled Water | 8 |

Stir until dissolved and then add the following chemicals in the order given, stirring each until dissolved:

| | | |
|---|---|---|
| *b* | Warm Distilled Water | 2½ |
| | Pure Sodium Sulfite | ½ |
| | Pure Acetic Acid (28%) | 1½ |
| | Potassium Alum Powder | ½ |

Add *b* to *a* and store in dark bottles away from light.

CHAPTER II

# ADHESIVES

### Packing Adhesive with Good Low-Temperature Performance

| | |
|---|---|
| "Butvar" B-76 | 12.0 |
| "Santicizer" B-16 | 8.0 |
| "Castorwax" | 444.0 |
| "Armid" HT | 0.9 |
| RA-50 ($TiO_2$) | 1.9 |
| Al-Sil-Ate W (Southern Clays) | 10.4 |
| "Polypale" ester 1 | 14.4 |
| "Staybelite" ester 10 | 10.6 |

Properties: Viscosity, 2000 cps at 150°C, (Brookfield, 3 at 12 rpm). Cream colored, nontacky film.

---

### Corrugated Case Pealer

| | |
|---|---|
| Water | 52.0 |
| Clay (china) | 21.5 |
| "Amaizo" 1108 Dextrin | 12.5 |
| "Amaizo" 1402 Dextrin | 5.0 |
| Corn syrup | 3.5 |
| Borax (Deca) | 2.8 |
| Caustic (15% solution) | 2.4 |
| Preservative | 0.2 |
| Defoamer | 0.1 |

Cook to 200°F, hold 15 min, cool.

---

### Pressure-sensitive Adhesive

| | |
|---|---|
| #1 Pale crepe | 100 |
| Zinc oxide | 50 |
| Aluminum hydrate | 50 |
| USP white oil | 50 |
| Hexane | 600 |
| "Wing-Stay" L antioxidant | 1 |
| "Wing-Tack" 95 resin | 70 to 90 |

---

### Hot Melt Adhesive

No. 1

| | |
|---|---|
| "Wing-Tack" resin | 100.0 |
| Polyethylene resin | 100.0 |
| Safflower oil | 22.0 |

No. 2

| | |
|---|---|
| Ethylene-vinyl acetate copolymer 27 to 29 | 33.0 |
| % vinyl acetate; melt index, 12-18 g/10 min | 33.0 |
| Refined paraffin wax; melting point 151°F | 33.0 |
| Rosin ester | 33.0 |
| "Irganox" 1010 | 0.5-1 |

---

**Dry Tape Adhesive, High-Temperature**

80 phr "Dexsil" 202
20 phr "Dexsil" 126 or 151
1 phr RO 3097 iron oxide
5 phr $HBO_2$
1 phr benxoyl peroxide
25 phr xylene

Dissolve the peroxide in warm xylene. Add hot "Dexsil" 202 (100-150°C) with continued heating and stirring. Add remaining ingredients in the following order: "DEXSIL" 126 or 151, iron oxide, and $HBO_2$. This mixture is then stirred and slowly heated removing the xylene until becoming adhesive or tacky. It is then allowed to stand overnight, approximately 12 hr, and then further heated until becoming rubbery (appears set or cured.) The stock can then be banded on a rubber mill and worked so as to obtain a good dispersion. Milling to desired thickness is the last step. Curing, in a circulating air oven should be stepwise as follows: 10 min at 100°F, 15 min at 300°, 30-60 min at 500°.

This curing period can vary, depending on the bonded surface, and as such, parts should be inspected frequently.

Resulting sheets should be relatively tack free and easily handleable.

Depending on the particular application, the adhesive can either, be used in past form, as a finished sheet, or with differing fillers (silicas, magnesium oxide and aluminium oxide).

Will withstand 700°F for 500 hours.

---

**Hot Pickup Label Adhesive**

| | | |
|---|---|---|
| "Butvar" B-76 | | 10.0 |
| Santicizer B-16 | | 10.0 |
| "Castorwax" | | 10.0 |
| "Armid" HT | | 0.2 |
| "Polypale" Ester | 1 | 51.5 |
| "Staybelite" Ester | 10 | 38.5 |

Properties: Viscosity, 2600 cps at 150°C, (Brookfield #3 at 30 rpm). Light yellow, tacky film.

**Adhesive Resin**

| | No. 1 Chemical-Resistant | No. 2 Heat-Sensitive | No. 3 Pressure-Sensitive |
|---|---|---|---|
| "Parlon" S125 | 20 | 12 | 8 |
| "Aroclor" 1254 | 6 | — | — |
| "Aroclor" 1260 | 6 | — | — |
| "Staybelite" Ester 3 | — | 20 | — |
| "Staybelite" Ester 10 | — | — | 12 |
| Dibutyl phthalate | — | — | 12 |
| Toluene | 68 | 68 | — |
| Acetone | — | — | 48 |
| Hexane | — | — | 20 |

**Wax Paper Adhesive**

| | |
|---|---|
| "Butvar" B-76 | 15.0 |
| "Santicizer" B-16 | 5.0 |
| "Castorwax" | 36.0 |
| "Armid" HT | 0.7 |
| "Acrawax" C | 2.0 |
| "Polypale" ester 1 | 24.6 |
| "Staybelite" ester 10 | 18.4 |

**Tape Adhesive**

| | A | B |
|---|---|---|
| "Ameripol" 4502 Crumb | 25.00 | — |
| "Ameripol" 1013 Crumb | 75.00 | 100.00 |
| "Pentalyn" H | 57.00 | 57.00 |
| "Staybelite" 10 | 57.00 | 57.00 |
| Zinc oxide | 40.00 | 40.00 |
| Titanium dioxide | 20.00 | 20.00 |
| Toluene | 315.00 | 315.00 |
| Antioxidant 2246 | 5.75 | — |

**Carpet to Concrete or Wood Adhesive**

| | |
|---|---|
| "Dow A" Antifoam | 0.33 |
| Ammonium oleate | 72.00 |
| Triethanolamine | 2.39 |
| Ammonium hydroxide | 5.30 |
| "Picco" 6070-3 | 428.00 |

| | |
|---|---|
| Water | 93.17 |
| McNamee clay | 150.00 |
| "Agerite" spar | 1.00 |
| "Marmix" 22332 | 173.50 |
| Natural latex 2XB (low-ammonia) | 15.40 |
| "Ultrawet" DS | 0.60 |
| "Acrysol" HV-1 | 30.00 |

Blend the first components, then add to "Picco" 6070-3 under high shear agitation. Add the water and clay slowly to obtain a uniform dispersion. Add the "Agerite" spar, then reduce the agitation while adding the "Marmix" 22332 and natural latex. The "Ultrawet" DS is added to the natural latex before addition to the compound. The "Acrysol" is added to increase the viscosity. Alternately, "Alcogum" 6625 may be used.

---

**Leather-to-Wood Adhesive**

| | |
|---|---|
| "Butvar" B-76 | 20.0 |
| "Santicizer" 141 | 2.5 |
| Rosin WW | 60.0 |
| "Castorwax" | 25.0 |
| "Armid" HT | 0.5 |

---

**Permanent Putty**

| | |
|---|---|
| "THIOKOL" FA (plasticity grade L) | 100.0 |
| MBTS | 0.4 |
| DPG | 0.1 |
| York white | 225-250 |
| Stearic acid | 0.25 |
| "Cumar" P-25 | 15.0 |
| *Additional MBTS | 2.1 |

Knife-grade putties are produced. They remain permanently pliable and soft and resist temperatures from —40 to above 200°F. In general they possess excellent adhesion to wood, metal, glass, plastics, and paints. They are used to seal and fill joints in storage bolted tanks for gasoline, oil, and various other solvents. Among other uses, they have also been used to seal pressurized cabins and fuel tanks on aircraft.

The above putty is an off-white color. With proper addition of colorants, colors from white to black can be mixed. Due to use of a large quantity of the cheap filler (calcium carbonate), the materials cost is very modest.

Mixing techniques are available to mix these putties on a mill without sticking, with the final reduction to a putty consistency by heating at 158 to 200°F. Or, with a hot internal mixer, the whole operation can be accomplished in one step.

---

**Window Putty**

| | |
|---|---|
| "Ameripol" 1009 crumb | 50 |
| "Ameripol" 1006 crumb | 50 |

| | | | |
|---|---|---|---|
| Heptane | 200 | "Calcene" TM | 160 |
| "Cumar" MH | 95 | | |

## Glass, Aluminum, Steel Construction Adhesives

| | No. 1 | No. 2 | No. 3 | No. 4 |
|---|---|---|---|---|
| "Thiokol" LP-32 | 100.00 | 100.00 | 100.00 | 100.00 |
| Stearic acid | 0.25 | 0.25 | 0.25 | 0.25 |
| "Witcarb" RC | 25.00 | 25.00 | 25.00 | 25.00 |
| "Titanox" RA-50 | 10.00 | 10.00 | 10.00 | 10.00 |
| "Aroclor" 1248 | 10.00 | 10.00 | 10.00 | 10.00 |
| "Durex" 10694 | — | 5.00 | — | 3.00 |
| "Tipox" B epxy resin | — | — | 5.00 | 3.00 |
| C-5 accelerator | 15.00 | 15.00 | 15.00 | 15.00 |

## Fire Retardant Mastic for Steel

Charge into mixer under agitation:

| | |
|---|---|
| Water | 280.0 |
| Potassium tripolyphosphate | 4.0 |
| "Phos-Check" P/30 | 235.0 |
| Melamine | 90.0 |
| Dipentaerythritol | 85.0 |
| "Chlorowax" 70 | 45.0 |
| "Resyn" 5000 | 26.0 |
| "Igepal" CTA-639 | 3.5 |
| Alkyd 100% N.V. | 15.0 |
| 6% cobalt drier | 0.5 |
| 24% lead drier | 1.0 |
| Asbestos S 33/65 | 50.0 |

## Panel Adhesive

| | |
|---|---|
| "Ameripol" 1013 crumb | 100 |
| "Cumal" MH | 25 |
| "Pentalyn" H | 25 |
| "Dixie Clay" | 5 |
| "Amberol" ST 137X | 15 |
| "Primol" 355 | 15 |

## Sealant, Non-sagging

PART A

| | |
|---|---|
| Polysulfide resin | 100.0 |
| Calcium carbonate (wet ground) | 20.0 |
| Titanium dioxide | 10.0 |
| Sulfur | 0.1 |
| Stearic acid | 1.0 |
| "Cab-O-Sil" M-5 | |
| {Ethylene glycol / "Cab-O-Sil" M-5} mix together | 1.2 |
| Ethylene glycol | 0.3 |
| "Triton" X-100 | 0.2 |

PART B

| | | |
|---|---|---|
| Accelerator (lead peroxide in DBT) | mix | 15.0 |
| Epoxy tackifier | together | 5.0 |

A nonsagging polysulfide sealant containing a phenolic tackifier can be prepared with 1.2 phr "Cab-o-sil" M-5 and 0.3% ethylene glycol dispersed in the resin. Excellent shelf stability and nonsagging properties during pot life can be obtained with this formulation. Since phenolic tackifiers stain masonry, epoxy tackifiers or other substitutes are often used.

When an epoxy tackifier is used in the formulation, it is added to the accelerator portion. In this formulation 1.6 phr "Cab-o-sil" M-5, 0.2% ethylene glycol, and 0.2% "Triton" X-100 is needed to produce a shelf stable sealant composition.

---

**Plastisol Automotive Sealer**

| | |
|---|---|
| Bakelite vinyl dispersion resin QYLF-2 | 24.54 |
| Bakelite vinyl resin solution VMCC | 3.92 |
| "Elvacite" 2044 | 0.98 |
| "Hycar" 1411 | 0.37 |
| "Camel White" | 29.20 |
| "Flexol" plasticizer 10-10 | 29.20 |
| "Flexol" plasticizer TCP | 2.70 |
| "Monomer X-970" | 4.91 |
| "Mark" LL | 0.49 |
| Resin-grade 244 asbestos | 3.68 |
| "Dicup" R | 0.01 |

Dissolve bakelite vinyl resin solution VMCC in "Flexol" plasticizer 10-10 at about 100°C. Dissolve "Elvacite" 2044 in 'Flexol" plasticizer TCP at 100°C. Mix the two solutions and all the other ingredients together in a pony, "Hobart," or similar mixer.

---

**"Styrofoam" to Plywood or Concrete Adhesive**

| | |
|---|---|
| "Picco" 6070-3 | 106.6 |
| "Piccolyte" S-10 | 15.0 |
| Mineral spirits | 10.0 |
| Ammonium oleate | 16.0 |
| Water | 15.0 |
| Acid casein | 6.7 |
| "Agerite" spar | 1.0 |
| McNamee clay | 100.0 |

| | |
|---|---|
| "Marmix" 21333 | 173.5 |
| Natural latex X2B (low-ammonia) | 15.4 |
| "Ultrawet" DS | 0.6 |
| ASE 60 | q.s. |

The emulsion is prepared using high shear agitation and the other components are blended in. After the McNamee clay has been added, the shear is reduced as the "Marmix" 21333 and natural latex are added. The "Ultrawet" DS is added to the natural latex before addtion to the compound. Viscosity is controlled by the judicious addition of the ASE 60.

---

**Vinyl Foam or Tile to Plywood-Concrete Adhesive**

| | | |
|---|---|---|
| | "Picco" 6070-3 | 145.0 |
| | Mineral spirits | 5.0 |
| Mix 1 | "Vistac" P | 10.0 |
| | Ammonium oleate | 5.0 |
| | "Antara" BL0344 | 24.0 |
| | "Antara" BL0344 | 3.0 |
| | Water | 50.0 |
| | "Nopco" 2271 | 15.0 |
| | "Rez-O-Sperse" | 22.6 |
| Mix 2 | Colloid kaolin | 150.0 |
| | Acid casein | 24.0 |
| | "Agerite" spar | 1.0 |
| | Sodium silicate 34 | 10.6 |
| | "Marmix" 22334 | 192.0 |
| | Natural latex 2XB (low ammonium) | 25.0 |
| | "Ultrawet" DS | 0.6 |
| | "Alcogum" 6625 | q.s. |

The materials in Mix 1 and 2 are mixed separately using high shear. They are blended together and with the "Marmix" 22334 and natural latex using moderate to low shear agitation. The "Ultrawet" DS is added to the natural latex before addition to the compound. The "Alcogum" is added as required, to obtain a paste viscosity.

---

**Durable Exterior Caulking Compound-White**

| | |
|---|---|
| Special T-blown soybean oil | 420 |
| Mineral spirits | 50 |
| Asbestos fiber | 50 |
| 1-5-Micron calcium carbonate | 350 |

| | |
|---|---|
| Talc | 450 |
| 30% titanium-calcium | 50 |
| 6% cobalt naphthenate (0.015%) | 1 |

Gunning may be improved at sacrifice in nonpenetration by the addition of up to 8 lb of free fatty acids.

For gray or natural remove titanium calcium pigment and replace with Ca $Co^3$.

Slump and crack resistance and/or gunning may be improved with sacrifice in nonpenetration and color of the white with additional asbestos fiber.

Lower raw material cost may be achieved by decreasing vehicle solids to 80% with mineral spirits at slight sacrifice in shrink resistance and nonpenetration.

---

**Caulk**

No. 1

| | |
|---|---|
| "Ameripol" 1013 Crumb | 100.00 |
| "Calcene" TM | 350.00 |
| "Wingstay" T | 1.00 |
| Linseed Oil | 81.00 |
| Wood Turpentine | 53.00 |

| | No. 2 | No.3 |
|---|---|---|
| "Ameripol" 1006 crumb | 25.00 | 100.00 |
| "Ameripol" 1009 crumb | 75.00 | — |
| "Wingstay" T | 1.00 | — |
| "AgeRite" resin D | — | 2.00 |
| Primol oil | 35.00 | — |
| "Sunthene" 380 | — | 5.00 |
| Cottonseed fatty acid | — | 10.00 |
| "Atomite" whiting | 150.00 | — |
| "Paragon" clay | — | 36.00 |
| "Calcene" TM | 75.00 | — |
| "Asbestine" 425 | 5.00 | — |
| "Catalin" 8318 | — | 23.00 |
| "Staybelite" ester | 20.00 | — |
| Titanium dioxide | 5.00 | — |
| Toluene | 75.00 | — |
| Hexane | — | 145.00 |
| Xylene | 50.00 | — |

| | | |
|---|---|---|
| Heptane | 75.00 | — |
| Alcohol | 1.00 | — |

No. 4

| | |
|---|---|
| "Acryloid" CS-1 | 48.50 |
| Pine Oil | 0.50 |
| "Ircogel" 2354 | 1.33 |
| Ethylene glycol | 1.25 |
| Calcium carbonate | 37.50 |
| Magnesium silicate | 8.50 |
| Asbestos | 1.50 |

No. 5

| | |
|---|---|
| "Ircogel" 2354 | 0.880 |
| "Loxite" 8017 | 35.210 |
| "Piccopale" 100 | 6.680 |
| Rosin | 0.775 |
| Titanium dioxide | 4.570 |
| Calcium carbonate | 36.260 |
| Magnesium silicate | 9.080 |
| Mineral spirits or Stoddard solvent | 5.170 |

Using a heavy duty mixer, add hydrocarbon resin solution, rosin, thixotropic agent into mixer. Mix until a smooth, heavy paste is obtained.

Add half of buty-rubber solution, the titanium dioxide, calcium carbonate, and magnesium silicate in that order, mixing until smooth and homogeneous.

Add remaining half of butyl-rubber solution, mixing well.

Add Stoddard solvent in increments, mixing well.

No. 6

(Gun and Squeeze-Tube-Grade Consistency)

| | |
|---|---|
| "Rhoplex" LC-40 | 430.17 |
| "Triton" X-405 | 9.46 |
| "Calgon T | 10.65 |
| "Paraplex" WP-1 | 124.21 |
| "Varsol" 1 | 26.91 |
| "Tamol" 850 | 1.27 |
| "Atomite" | 692.06 |
| "Ti-Pure" R-901 | 17.72 |

To prepare caulks and sealants based on Rhoplex LC-40, closed sigma-blade, "Ross," "Pony Mixer," "Ribbon Blender," or similar type high-shear, low-speed mixing equipments are recommended. The caulk ingredients should be added slowly to the mixer in the order listed. Addition of the "Triton" X-405 immediately after the "Rhoplex" LC-40 is desirable to stabilize the emulsion and prevent excessive thickening during addition of the plasticizer and extender. After addition of the last ingredient, the lid of the mixer should be tightly sealed to prevent water loss. Total mixing time of about one and a half hours is recommended.

---

**Moisture-Cure Urethane Sealant**

| | |
|---|---|
| "XP"-1387 | 575 |
| Toluol | 14 |
| Carbon black (thermal type) | 430 |
| Silica aerogel | 43 |

Self-leveling grades can be made by replacing a minor portion of the carbon black with calcium carbonate and eliminating the silica from the formulation.

Amino-silanes may be added for improved adhesion at 1-2% of the vehicle.

---

**Automobile Body Sealer**

| | |
|---|---|
| 240°F anuline point petroleum oil | 40 |
| "Barber" gilsonite sparkling black pulverized fine grind | 33 |
| China clay | 14 |
| Calcium carbonate | 10 |
| Asbestos fiber 7RF | 2 |
| "Armeen" | 1 |

---

**Tile Adhesive**

| | A | B |
|---|---|---|
| "Ameripol" 1013 crumb | 25.00 | 25.00 |
| "Ameripol" 1009 crumb | 75.00 | 75.00 |
| "Calcene" TM | 40.00 | 40.00 |
| "Paragon" clay | 120.00 | 120.00 |
| "Cumar" MH 2½ | 95.00 | — |
| "Staybelite" ester 10 | — | 95.00 |
| Antioxidant 2246 | 1.00 | 1.00 |
| Heptane | 175.00 | 175.00 |

| | C |
|---|---|
| Rosin | 17.0 |
| Furfural | 2.0 |
| Drying oil | 6.8 |
| White spirit | 9.8 |
| Acetone | 0.9 |
| Dolomite (powder) | 63-70 |

Dissolve the ground rosin and furfural in white spirit at 80°C. Add oil and stir to a homogeneous mass. Cool and add the acetone adding the powdered dolomite slowly with continuous stirring to a workable consistency.

---

**"Neoprene" Cement**

No. 1

| | | |
|---|---|---|
| A. | "Neoprene" Rubber | 100.0 |
| | "Neozone" A | 2.0 |
| | Zinc Oxide | 5.0 |
| | Magnesium Oxide | 4.0 |

Milling Procedure

Band "Neoprene" rubber on cold rolls forming a continuous sheet. Add premixed Neozone A, zinc oxide and magnesium oxide. Mill for two minutes. Sheet off.

| | | |
|---|---|---|
| B. | Solvent or Solvent Blend | 100.1 |
| | Water | 1.0 |
| | Magnesium Oxide | 5.0-7.2 |
| | "Durez" 26228 Resin | 45.0 |
| | Pre-react for two hours, add | |
| C. | Solvent or Solvent Blend | 543.0 |
| | Rubber Recipe | 111.0 |

The adhesion of "Neoprene" cements to porous and nonporous surfaces is increased when "Durez" 26228 is utilized in the cement formulation. The addition of 10 to 50 phr of 26228 resin will produce an increase in adhesion to porous substrates such as wood, paper, etc.; to synthetic fibers such as "Dacron," nylon and rayon; also to nonporous substrates such as glass, metals and unplasticized plastic films. The addition of more than 50 phr of resin in the formulatoin will continue to improve adhesion at a sacrifice of adhesive film flexibility. A marked increase in heat resistance is also obtained as the resin content in the formulation is increased.

No. 2

| | |
|---|---|
| "Neoprene" AC | 100.0 |
| Magnesium Oxide | 10.0 |
| Zinc oxide | 5.0 |
| Antioxidant | 2.0 |
| Water | 0.5 |
| Heat-reactive phenolic | 45.0 |
| "Nirez" 2092 | 60.0 |
| Solvent (1:1 Toluene, MEK) | 666.0 |

---

### Asphalt Mastic

| | |
|---|---|
| **Component A** | |
| Gilsonite selects pulverized | 46 |
| Calcium carbonate or silica (100-mesh) | 46 |
| Asbestos 7r | 8 |
| **Component B** | |
| Asphalt—100 penetration | 38 |
| Gilsonite selects (lump or pulverized) | 17 |
| Mineral spirits | 23 |
| Xylol | 22 |

**Mix ratio:** 100 parts component A
70 parts component B

This mixture will be of heavy consistency suitable for trowel application, such as for floor patching. It will have a workable pot life of about 30 min and will cure hard in 24 hr at ambient temperatures.

Each component may be varied to achieve specific properties. Component B can include such materials as rubber, oils, or resins for special applications. Similarly, Component A can include ground cork (insulation) or other fillers for specific uses.

Chapter III

# CEMENT & RELATED PRODUCTS

**Concrete Block Filler**

| | |
|---|---|
| Water | 430 |
| "Asbestine" 325 | 382 |
| "Duramite" | 190 |
| "Ti-Pure" R-902 | 29 |
| Asbestos 7RF-1 | 19 |
| Emulsion (51.5% "Genepoxy" M195 in $H_2O$) | 110 |
| "Versamid" 265-WR70 | |

Stir the pigments and filler into the water and continue stirring with low shear equipment for a few minutes. The epoxy emulsion is added last to complete this package. Curing agent is added and mixed by hand to provide the complete product. Material cost is competitive with latex based block fillers. A 55% PVC filler can be made by adding the following to the above formula:

---

**Concrete Curing Membrane**

No. 1

| | |
|---|---|
| "Pliolite" S-5D | 15.0 |
| Mineral Spirits | 42.5 |
| "Solvesso" 100 | 42.5 |

No. 2

| | |
|---|---|
| "Pliolite" ACL | 12.0 |
| Xylene | 88.0 |

---

**Release Agent for Hot Concrete**

No. 1

| | |
|---|---|
| "FT"-100H or FT-200 | 30.5 |
| Paraffin 125/127 AMP | 3.5 |
| Emulsifier AP-6 | 6.0 |
| Triethanolamine | 2.5 |
| Water | 57.5 |

**Lb 21-130 Solvent-based Release Agent**

No. 2

| | |
|---|---|
| "Duroxon" E-421R | 2.0 |
| "Duron" 180D | 2.0 |
| Paraffin 133/135°F AMP | 4.0 |
| Mineral spirits | 92.0 |

---

**Masonry Surface Filler**

No. 1

| | |
|---|---|
| "Pliolite" S-5A | 62.00 |
| Chlorinated paraffin 40% | 33.00 |
| Mineral spirits No. 10 | 65.00 |
| "Solvesso" 100 | 65.00 |
| "Nuosperse" 657 | 8.00 |

Mix until resin is completely dissolved and add slowly:

| | |
|---|---|
| Asbestos shorts 7R | 50.00 |
| Superwhite silica | 400.00 |
| Calcium carbonate | 200.00 |
| "Bentone" 38 | 5.00 |

Add sufficient solvent blend to prevent balling up of pigments, but do not over thin:

| | |
|---|---|
| Methanol | 1.50 |
| Mineral spirits No. 10 | 134.00 |
| "Solvesso" 100 | 134.00 |

No. 2

| | | |
|---|---|---|
| Pre-mix to form | "Pliolite AC-3 | 44 |
| gel. To the AC-3 in a | "Varsol" 18 | 266 |
| Cowles dissolver add: | "Pliolite" VT | 44 |
| After the mixture | | |
| is dissolved, add: | "Titanox" RCHT | 133 |
| | "Gold Bond "R" | 170 |
| | White Portland cement | 243 |
| | "Varsol" 18 | 323 |
| | | 1,223 |

Pony mixer procedure for producing Pliolite based sealer.

| | |
|---|---|
| Into a pony mixer add: | "Pliolite" AC-3, |
| | "Varsol" 18. |
| Mix for appr. 2½-3 hr, then add: | "Pliolite" VT-L, |

| | |
|---|---|
| | "Varsol" 18. |
| Dissolve for appr. 1 hr, then add: | "Titanox" RCHT, "Gold Bond" "R", White Portland cement. |

**Improving Plaster Molding**

In molded plaster products, "Lignosol" SF or SFX at rates of 0.25 to 1.0% based on solids permits reduction of water by approximately 20 to 25%. Better green strengths and dry strengths result. Drying is accelerated.

**Asphalt Curbing**

| | |
|---|---|
| Aggregate | 9300 |
| Paving grade asphalt, 85-100 pen. | 665 |
| Pulverized gilsonite | 35 |

*Note:* Based on the chart, the addition of 5% gilsonite to an 85-100 pen. asphalt cement will reduce the penetration to 50-60 pen.

The heated aggregate and pulverized gilsonite is added to the pug mill first, followed by the required amount of 85-100 pen. asphalt. The laying requirements, i.e. transportation life, of the curbing mix is about the same as for a typical hot mix paving material.

**High-Temperature Ceramic Foams**

They consist basically of zircon suspended in a fine bubble structure. Composition of typical mixes follow the range, in percent by weight:

| | |
|---|---|
| Refractory oxide | 54.00—75.0 |
| Water | 8.00—16.0 |
| Silicate binder | 14.00—30.0 |
| Clay | 0.00— 2.0 |
| Whipping agent | 0.15— 1.0 |
| Sodium silicofluoride | 1.00— 3.0 |
| Fibrous material | 1.50—10.0 |

Density of the material can be predicted and adjusted by changing composition as well as foaming time.

CHAPTER IV

# COATINGS

### Basement Wall Sealer

No. 1

Into a Cowles dissolver premix:

| | |
|---|---|
| "Pliolite" AC-3 | 45 |
| "Varsol" 18 | 270 |

After mixing for approximately 45 minutes, add:

| | |
|---|---|
| "Pliolite" VT | 45 |

After the mixture is dissolved, add:

| | |
|---|---|
| "Titanox" RCHT | 172 |
| "Cold Bond" R | 247 |
| White portland cement | 319 |
| "Varsol" 18 | 135 |

RCM, $0.90/gal; viscosity, 115 KU; coverage, 75 to 125 sq.ft/gal.

2% Cal Ink 6600 series colorant can be added for coloring based on the total weight of the paint.

No. 2

| | |
|---|---|
| "Pliolite" VT | 87 |
| "Varsol" 18 | 203 |

After dissolving, add:

| | |
|---|---|
| "Titanox" RCHT | 167 |
| "Cold Bond" R | 239 |
| White portland cement | 318 |

After temperature reaches 120° F, add:

| | |
|---|---|
| "Thixatrol" ST | 11 |

Finally reduce with

| | |
|---|---|
| "Varsol" 18 | 189 |

---

### Masonry Water Proofing

Mix 1 gal Dow Corning 772 water repellent as received with 11 gal water. Do not use concentrations higher than 3% solids. Dilute solutions of the siliconate can be applied by many methods, including dipping, spraying and brushing.

**Application Precautions:** Under certain conditions a white precipitate of calcium carbonate may be deposited as the water repellent dries. To avoid this, which is especially noticeable on dark surfaces, observe the following precautions: (1) Avoid rundown; avoid overlapping or second coating; (2) Do not allow water repellent to form puddles on flat surfaces; (3) Do not allow water repellent to contact glass or aluminum. If contamination occurs, wash thoroughly before water repellent dries to avoid staining or etching.

After application of the silicone solution, the treated surface should be allowed to dry at least 24 hr to develop maximum water repellency. This interval may be shortened somewhat by force-drying at temperatures to 300°F. Though this removes the water quickly, time must still be allowed for the curing reaction.

---

**Masonry Paints**

| | For Swimming Pools | For Bricks, Asbestos Shingles, and Similar Surfaces | For Porous Masonry | |
|---|---|---|---|---|
| | No. 1 | No. 2 | No. 3 | No. 4 |
| "Parlon" S10 | — | — | 5.30 | — |
| "Parlon" S20 | 8.00 | 11.50 | — | 8.60 |
| "Duraplex' C-49 | 4.00 | — | — | — |
| "Clorafin" 40 | — | 4.35 | 1.95 | — |
| "Aroclor" 1254 | 4.00 | — | — | — |
| "Duraplex" D-65A (85%) | — | 7.11 | 3.27 | — |
| Raw linseed oil | — | — | — | 6.40 |
| Linseed oil, heat-bodied, viscosity Z3 | — | — | — | 6.40 |
| Titanium dioxide | 15.20 | 10.30 | 20.60 | 11.40 |
| Zinc oxide | — | 3.90 | 4.10 | 4.30 |
| "Asbestine" 3X | 15.20 | 6.40 | 10.40 | 7.20 |
| Mica (325 mesh white waterground) | — | 5.10 | 6.30 | 5.80 |
| "Bentone" 34 | 1.60 | — | 2.50 | 2.90 |
| "Amsco Solv" D | — | 33.74 | 31.46 | 31.16 |
| "Solvesso" 100 | — | 13.40 | 10.60 | 12.00 |
| Turpentine | — | 3.54 | 3.20 | 3.30 |
| Xylene | 51.96 | — | — | — |
| Epichlorohydrin | 0.04 | 0.06 | 0.03 | 0.04 |
| "ERL" 2774 | — | 0.60 | — | 0.40 |
| Cobalt naphthenate, 6% | — | — | 0.03 | 0.10 |
| "Dyphos" | — | — | 0.26 | — |

Light-blue or green pastel colors are popular for swimming pools. Formula 1 can be tinted to obtain suitable colors. Alkali-resistant pigments such as phthalocyanine blues or greens should be used for tinting.

Formula 1 is also useful as a starting formulation to meet specification TT-P-95. This specification covers a self-priming paint for interior or exterior use on concrete swimming pools.

In applying paints to swimming pools, cracked or scaling paint should be removed from the surface to be painted. Fresh concrete should be acid-etched or wire-brushed before painting. The paint should have time to dry thoroughly before filling the pool.

Formula 2, is suggested for application to nonporous stucco, brick, and asbestos shingles, as both an interior and an exterior coating. This paint has sufficient alkali resistance for use on the above surfaces and adequate salt-fog resistance for seacoast exposures.

---

## Emulsion-Type Terrazzo Paint (Nondiscoloring)

| | |
|---|---|
| Part I | |
| "Efton" D Super | 6.6 |
| "Duroxon" R-21 | 5.4 |
| "Durmont" E | 6.4 |
| "Aristowax" 165 | 5.6 |
| Stearic acid, double-pressed | 5.0 |
| Mineral spirits | 15.0 |
| Part II | |
| Triethanolamine | 3.0 |
| Borax | 1.0 |
| Water 210°F | 89.0 |

Melt the waxes together with the stearic acid and slowly add the mineral spirit. If the solution is cloudy, heat the batch slightly until it is clear. In a separate vessel dissolve the triethanolamine and borax in the indicated amount of water. Adjust the temperature to approximately 190 to 205°F, start high-speed agitation, and slowly but steadily pour I into II. Cool with agitation to approx. 100°F max. Store the product in a covered container.

| Concrete Floor Paint | No. 1 | No. 2 | No. 3 | No. 4 |
|---|---|---|---|---|
| "Parlon" S20 | 13.90 | 6.50 | 17.80 | — |
| "Parlon" S125 | — | 6.50 | — | — |

| | | | | |
|---|---|---|---|---|
| "Parlon" S10 | — | — | — | 15.90 |
| "Duraplex" D-65A (85% solids) | 8.40 | — | — | — |
| "Pentalyn" 802A varnish* | — | 11.50 | — | — |
| "Aroclor" 1254 | — | 4.30 | 9.90 | — |
| "Clorafin" 40 | 5.20 | — | — | 4.95 |
| "Duraplex C-49 (100% solids) | — | — | — | 5.55 |
| "Aroclor" 5460 | — | — | 5.90 | — |
| Titanium dioxide | 17.30 | — | 15.40 | 16.20 |
| Zinc oxide | — | — | 1.90 | 3.20 |
| Carbon black | — | -- | 0.50 | — |
| "Asbestine" 3X | — | — | — | 2.20 |
| Lampblack | 0.13 | — | — | — |
| Red iron oxide | — | 13.80 | — | — |
| "Amsco" solv D | 36.70 | — | — | — |
| Mineral spirits | — | 11.29 | 9.80 | 10.42 |
| "Solvesso" 100 | 14.11 | 45.30 | 38.75 | 41.50 |
| Turpentine | 3.80 | — | — | — |
| Cobalt naphthenate, 6% | 0.05 | 0.09 | — | 0.03 |
| Manganese naphthenate, 6% | — | 0.09 | — | — |
| Zinc naphthenate, 8% | — | 0.43 | — | — |
| Epichlorohydrin | 0.04 | 0.04 | 0.05 | 0.05 |
| "Eskin" No. 1 | 0.03 | 0.03 | — | — |
| Soybean lecithin | 0.34 | 0.13 | — | — |

*Preparation of Pentalyn 802A varnish.

| | |
|---|---|
| Pentalyn 802A | 100 lb |
| Tung oil | 20 gal |
| Linseed oil, alkali-refined | 20 gal |

Heat the "Pentalyn" 802A and 10 gal of linseed oil together at 585°F (for 1 hr to 1 hr 20 min). Add china wood oil; temperature drops to 370°F ± 10°F. Reheat to 575°F as fast as possible (40 min maximum), and hold for heavy drop from hot stirrer (0 to min). Turn heat off and check with 10 gal of linseed oil. Air-cool to 350°F, and thin to 90% nonvolatile with xylene.

## Satin Sheen Urethane Concrete Floor Enamel

| | |
|---|---|
| "Santocel" FR-C | 10.0 |
| "Multron" resin | 18.0 |
| Solvent mixture* | 71.6 |
| Lecithin (50% in xylol) | 0.4 |

*"Cellosolve" Acetate (Urethane grade), Toluol, Xylol (1:2:1).

### Floor Sealer

| | No. 1<br>16% solids | No. 2<br>40% solids |
|---|---|---|
| "RWL"-201 latex (40%) | 340 | 850 |
| N-Methyl-2-pyrrolidone | 16 | 40 |
| "Igepal" CO-430 | 8 | 20 |
| Water | 636 | 90 |
| | 1000 | 1000 |
| Ammonium hydroxide (28%) | 2 | 2 |
| Formalin | 1.5 | 4 |

Dilute RWL-201 Latex with water and adjust to a pH of 7.0 to 7.5 with 28% ammonium hydroxide (Precaution: Do not raise pH of high-solids RWL-201 latex above 7.5). Add other ingredients and stir for ½ hr. The pH is adjusted to 8.0 to 8.5 with 28% ammonium hydroxide.

No. 2

| | |
|---|---|
| "Acryloid" B-66 | 27.00 |
| "Amberlac" 292X (48.5%) | 3.10 |
| Xylol or toluol | 69.90 |

No. 3

| | |
|---|---|
| U-2006 "Ubatol" | 70.0 |
| U-3054 "Ubatol" | 30.0 |
| Tributoxyethyl phosphate | 1.0 |
| 2-Pyrrolidone | 4.0 |
| Ethylene glycol | 1.0 |
| "FC"-134 1% | 1.0 |

No. 4

| | |
|---|---|
| "Rhoplex" B-60A | 16.3 |
| "Rhoplex" B-85 | 19.7 |
| Water | 61.0 |
| Ethylene glycol | 3.0 |

The "Rhoplex" B-60A and "Rhoplex" B-85 should be blended with thorough agitation and the required water introduced. After stirring for at least 5 min, the ethylene glycol should be added slowly with agitation. Stirring should be continued for at least min to ensure complete homogenization. The blend should be aged for a minimum of 24 hr before use.

The addition of 0.35 parts of "KP"-140 will improve the levelling and water-resistance of this formulation. The freeze-thaw stability is reduced to two cycles by this addition.

### Abrasion-Resistant Floor Finish

| | |
|---|---|
| *Solids* | |
| "Parlon" S20 | 50 |
| "Aroclor" 1254 | 25 |
| "Duraplex" C-55 | 25 |
| *Solvents* | |
| "Solvesso" 100 | 80 |
| Xylene | 10 |

| | |
|---|---|
| Turpentine | 10 |

**PVC Clear Topcoat**
**(Stain-resistant floor covering)**

| | |
|---|---|
| "Diamond" PVC-40 | 100 |
| Epoxidized soybean oil | 3 |
| Ba-Cd-Zn stabilizer | 3 |
| Plasticizer 466 | 26 |
| Optical brightener | as desired |

**Finish for Vinyl Asbestos Linoleum, and Asphalt Flooring**

| | |
|---|---|
| "Rhoplex" B-217 (12%) | 50.0 |
| "Poly-Em" 40 (12%) | 50.0 |
| Ethylene glycol | 1.0 |
| Tributoxyethyl phosphate | 0.8 |
| "Triton" N-101 | 0.4 |
| "FC-128" (1%) | 0.8 |

The components should be added in the order in which they are listed above. "Rhoplex" B-217 should be diluted before the incorporation of the "Poly-Em" 40. Before addition, the tributoxyethyl phosphate and ethylene glycol should be diluted with an equal amount of water and emulsified with a portion of the "Triton" N-101. The water for this dilution can be calculated as part of the water for dilution of the other components.

**Architectural High-Gloss Enamel**

| | |
|---|---|
| Rutile titanium dioxide | 300.0 |
| Zinc oxide | 25.0 |
| "Kelsol" DV-1695 | 515.0 |
| 6% cobalt naphthenate (0.06%) | 2.5 |
| 6% manganese naphthenate (0.02%) | 1.0 |
| 24% lead naphthenate (0.1%) | 1.0 |
| Bubble breaker-SK | 5.0 |
| Water | 250.0 |

**Tile-Like Enamel**

Component I (Pebble mill grind):

| | |
|---|---|
| Rutile titanium dioxide | 300.0 |
| Phthalo blue | 1.0 |
| CASTOR 1066 | 240.0 |
| "Cellosolve" acetate | 230.0 |
| "Kelecin" F | 6.0 |

Tumble overnight, adjust, filter, and package.

Component II

| | |
|---|---|
| "Spenkel" P49-75S | 335.0 |

Add Component II to Component I and mix thoroughly 10 to 20 min before use. For spray application, approximately half of the

"Cellosolve" acetate should be replaced with xylol and/or ethyl acetate. Pot life at least 1 working day.

---

**Quick-Drying Interior Trim Enamel—White**

| | |
|---|---|
| Rutile titanium dioxide | 300.0 |
| "Spenkel" F48-50MS | 600.0 |
| Mineral spirits | 60.0 |
| 6% cobalt naphthenate (0.03%) | 1.5 |
| 24% lead naphthenate (0.3%) | 3.8 |
| "Exkin" No. 2 (0.2%) | 0.6 |

---

**Air-Drying Green Enamel**

| | |
|---|---|
| "Stepanyl" 324-57 | 56.0 |
| Rutile $TiO_2$ (1) | 67.5 |
| Water Dispersed Green (26.1% N.V.) | 29.8 |
| Deionized $H_2O$ | 77.6 |
| n-Hexanol | 2.6 |

Charge pebble mill, grind 48 hr. Let down:

| | |
|---|---|
| "Stepanyl" 324-57 | 465.0 |
| 24% Pb drier | 0.2 |
| 6% Mn drier | 1.3 |
| Deionized water | 300.0 |

---

**Gray Baking Enamel**

| | |
|---|---|
| "Stepany" 324-57E | 165.0 |
| Barytes | 70.3 |
| Rutile $TiO_2$ | 77.0 |
| Carbon black | 2.7 |
| Aluminum silicate | 26.4 |
| Basic lead silico Chromate | 26.4 |
| "Cellosolve" | 6.4 |
| Deionized water | 194.3 |

Charge pebble mill, grind 22 hr. Let down:

| | |
|---|---|
| "Advacar" 6% Mn Drier | 2.2 |
| "Stepany" 324-57E | 238.0 |
| Deionized water | 107.5 |
| n-Hexanol | 5.5 |
| 10% solution, water-soluble silicone (6) | 3.1 |
| Water-soluble melamine (80% N.V.) (7) | 75.2 |

**Air-Drying Blue Enamel**

| | |
|---|---|
| "Stepanyl" 324-57 | 64.0 |
| Rutile $TiO_2$ | 77.0 |
| Blue Pigment | 8.6 |
| Deionized water | 89.0 |
| n-Hexanol | 3.0 |

Charge pebble mill, -grind 46 hr. Let down:

| | |
|---|---|
| "Stepanyl" 324-57 | 363.0 |
| 24% Pb (drier) | 0.2 |
| 6% Mn (drier) | 1.1 |
| Deionized water | 394.1 |

---

**White Baking Enamel**

| | |
|---|---|
| "Cabot" RF-30 | 140 |
| "Beetle" 227-8 | 52 |
| Butyl alcohol | 8 |

Premix and then disperse on a roll mill to a fineness of grind of 7 ns.

| | |
|---|---|
| "Beetle" 227-8 | 44 |
| "Rezyl" 7365-51 | 100 |
| "Rezy" 99-5 | 80 |
| VM&P naphtha | 56 |
| Xylol | 15 |
| Total weight | 495 lb |
| Total yiled (approx.) | 50 gal |

This coating should be reduced 4 to 1 with xylol for spray application. Bake at 325°F for 15 min.

---

**White Four-Hour Exterior Enamel, Lead Free**

| | |
|---|---|
| "Titanox" RC | 83 |
| "Titanox" R-610 | 250 |
| Aluminum stearate | 2 |
| Soya lecithin | 4 |
| "Aroplaz" 1085-M50 | 172 |

Premix and then disperse on a thre-roll mill to a fineness of grind of 7 ns.

| | |
|---|---|
| "Aroplaz" 1085 M50 | 490 |
| "Synthenol" G-H | 45 |
| No. 460 Solvent | 56 |
| Nuodex Cobalt Octoate 6% | 5 |
| Nuodex Calcium Naphthena e 4% | 13 |
| "Exkin" No. 2 | 2 |
| Total weight | 1122 lb |
| Total yield (approx.) | 110 gal |

This coating should be brushed full body.

**Grey Hammered Finish Enamel**

| | |
|---|---|
| 1593 "Albron" Aluminum Paste | 9 |
| "Cycopol" S-102-5 | 224 |
| "Uformite" MM-55 | 40 |
| VM&P naphtha | 148 |
| Triethanolamine | 1 |
| Total weight | 422 lb |
| Total yield (approx.) | 55 gal |

Carefully break up the aluminum paste in a portion of the "Cycopol" resin by slowly adding the resin to the aluminum while maintaining agitation. Once the aluminum paste is fully broken up and the mixture is homogeneous, the remaining ingredients are added to complete the batch. This coating should be sprayed full body. The baking schedule is 300°F, 15 min.

---

**White "Thumb Tack" Roller Coating Enamel**

| | |
|---|---|
| "Cabot" RF-30 | 150.0 |
| Aluminum stearate | 1.0 |
| "Uformite" MM-55 | 18.0 |
| "Rezyl" 99-5 | 60.0 |

Premix and then disperse on a three-roll mill to a fineness of grind of 7 ns.

| | |
|---|---|
| "Rezyl" 99-5 | 240.0 |
| "Uformite" MM-55 | 84.0 |
| Petrolatum | 1.0 |
| Nuodex cobalt naphthenate 6% | 0.3 |
| Pine oil | 5.0 |
| Triethanolamine | 1.0 |
| Total weight | 560.3 lb |
| Total yield (approx.) | 55.0 gal |

This enamel should be applied full body by roller. Bake at 300°F for 10 min.

---

**Air-Dry Gray Spray or Dip Enamel**

| | |
|---|---|
| "Stepanyl" 324-57 (50% n.v.) | 177.0 |
| Barytes | 18.9 |
| Rutile $TiO_2$ | 21.0 |
| Carbon black | 0.7 |
| Basic lead silicochromate | 14.2 |
| "Cellosolve" | 7.0 |

| | |
|---|---|
| Deionized water | 199.3 |
| Charge pebble mill, grind 22 hr. Let down: | |
| 6% Mn drier | 1.3 |
| 24% Pb drier | 0.2 |
| "Stepanyl" 324-57 | 257.0 |
| Deionized water | 295.4 |
| n-Hexyl alcohol | 4.7 |
| 10% solution, water-soluble silicone | 3.3 |

---

**Gloss Black Baking Dip or Spray Coat**

| | |
|---|---|
| "Stepanyl" 324-57E | 201.0 |
| Carbon black | 19.5 |
| Supermicrofine barytes | 32.5 |
| Basic silicate white lead | 6.5 |
| Deionized water | 200.0 |
| Pine oil | 4.9 |
| Charge pebble mill, grind 30 hr. Let down: | |
| 6% Mn drier | 2.4 |
| "Stepanyl" 324-57E | 268.0 |
| Water-soluble melamine resin (80% n.v.) (4) | 103.2 |
| Deionized water | 162.0 |

---

**Air Dry Gray Semi-Gloss**

| | |
|---|---|
| "Stepanyl" 324-57 | 183.00 |
| Natural microfine barytes | 78.00 |
| Rutile $TiO_2$ | 85.00 |
| Carbon black | 2.93 |
| Basic lead silicochromate | 32.00 |
| n-Butanol | 7.17 |
| Deionized water | 185.00 |
| Charge pebble mill, grind 24 hr. Led down: | |
| 6% Mn drier | 1.80 |
| 24% Pb drier | 0.60 |
| "Stepanyl" 324-57 | 264.50 |
| Deionized water | 149.50 |
| Water soluble silicone (10% n.v.) | 3.10 |

---

**One-Coat House Paint, White**

| | |
|---|---|
| Rutile titanium dioxide | 100.00 |

| | |
|---|---|
| Anatase titanium dioxide | 200.00 |
| 35% leaded zinc oxide | 400.00 |
| Micronized magnesium silicate | 150.00 |
| T-1215 Z2 Heat bodied linseed oil | 144.00 |
| A-R Varnish and grinding linseed oil | 266.00 |
| Mineral spirits | 149.00 |
| 24% lead naphthenate (0.4%) | 7.00 |
| 6% manganese naphthenate (0.03%) | 2.00 |

**Exterior Latex Paint**

**Mill Paste**

| | |
|---|---|
| "Ti-Pure" R-610, rutile titanium dioxide | 200.00 |
| "Ti-Pure" FF, anatase titanium dioxide | 50.00 |
| "Vi-Cron" 25-11, calcite | 121.80 |
| Mica, water ground, 325 mesh | 13.50 |
| "Tamol" 731 (25% solids) | 1.54 |
| Potassium tripolyphosphate (10% solids) | 15.40 |
| "Triton" CF-10 (25% solids) | 7.70 |
| "Balab" 648 bubble breaker or "Nopco" NXZ | 0.90 |
| Phenylmercuric acetate (6% solids) | 10.50 |
| 2% "Cellosize" WP-4400 | 50.00 |
| Water | 75.00 |

**Let down:**

| | |
|---|---|
| VV 10-VA latex (52% solids) | 433.50 |
| Phenylmercuric acetate (6% solids) | 10.50 |
| Butyl "Dioxitol" /ethylene glycol (1/1 by wt.) | 45.40 |
| 2% "Cellosize" WP-4400 | 96.60 |
| Water | 2.20 |
| Ammonium hydroxide to pH 7.6 | — |

**Styrene Butadiene Interior Latex Paint**

| | |
|---|---|
| Water | 453.0 |
| "Super AD-IT" | 1.0 |
| KTPP | 1.5 |
| Lecithin wd | 4.5 |
| "Nuodex" AF-100 | 3.0 |
| $TiO_2$ (rutile) | 125.0 |
| "Hydrite" regular | 250.0 |
| Calcined clay | 100.0 |
| "Cab-o-sil" M-5 | 5.0 |
| "Triton" X100 | 2.0 |
| "Gen-flo" 67 | 220.0 |
| "Methocel" | 5.0 |

**Polyvinyl Acetate Latex Interior Paint**

| | |
|---|---|
| Water | 325 |

| | |
|---|---|
| "Super AD-IT" | 1 |
| "Tamol" 731 | 6 |
| KTPP | 1 |
| Lecithin wd | 4 |
| "Nuodex" AF-100 | 3 |
| Ethylene glycol | 30 |
| $TiO_2$ (rutile) | 225 |
| Clay hydrite | 200 |
| Wet ground limestone | 100 |
| "Methocel" | 3 |
| "Resyn" 1251 | 310 |

**Acrylic Exterior Latex Paint**

| | |
|---|---|
| Water | 174.0 |
| "Super AD-IT" | 10.0 |
| "Natrosol" 250HR | 2.5 |
| "Nuodex" AF-100 | 3.0 |
| "Tamol" 731 | 11.0 |
| "Triton" CF10 | 2.0 |
| Ethylene glycol | 20.0 |
| $TiO_2$ (rutile) | 240.0 |
| $TiO_2$ (anatase) | 10.0 |
| "Nytal" 300 | 100.0 |
| "Duramite" | 90.0 |
| "Rhoplex" AC-34 | 512.0 |

**Classic Exterior House Paint White**

| | |
|---|---|
| Rutile titanium dioxide | 50 |
| Anatase titanium dioxide | 150 |
| 35% leaded zinc oxide | 400 |
| Talc | 225 |
| "Drisoy" Z2-Z3 | 144 |
| "A-R" Varnish and grinding linseed oil | 266 |
| Mineral spirits | 150 |
| 6% manganese naphthenate (0.03%) | 2 |
| 24% lead naphthenate (0.4%) | 7 |

**Scrub-Resistant White Flat Wall Paint**

| | |
|---|---|
| "TiPure" R-901 | 55 |
| "Satintone" No. 1 | 195 |
| "Camel" carb | 242 |
| Water | 210 |
| 25% Soln. "Tamol" 731 | 5 |
| "Nopco" NDW | 2 |
| "Troysan" CMP acetate | 2 |
| Ethylene glycol | 15 |

Grind with high-speed mixer:

| | |
|---|---|
| "Piccolloid" Emulsion No. 15 | 88 |
| "Flexbond" 315 | 83 |
| 2% solution "Cellosize" QP-4400 | 266 |

## Maintenance Paint Topcoats

| | Formula | | |
|---|---|---|---|
| | A | B | C |
| Chlorinated polyolefin 310-6 | 57.00 | 59.7 | 60.2 |
| "Elvax" 40 wax | — | 39.7 | 30.1 |
| "Aroclor" 1260 resin | — | — | 9.0 |
| "Aroclor" 1254 resin | 22.60 | — | — |
| "Aroclor" 5460 plasticizer | 13.75 | — | — |
| Stabilizer A-5 | 0.50 | 0.6 | 0.5 |
| Epichlorohydrin or propylene oxide | 0.50 | — | 0.2 |
| "KODAFLEX" DOS plasticizer | 5.65 | — | — |

## Low-Cost Shingle and Shake Paint

| | |
|---|---|
| Rutile titanium dioxide | 150.0 |
| Calcium carbonate | 400.0 |
| Diatomaceous silica | 75.0 |
| Aluminum stearate | 7.5 |
| Carbon black | 2.0 |
| "Cykelin" 70 | 320.0 |
| "Kelecin" F | 6.0 |
| Mineral spirits | 200.0 |
| 6% cobalt napthenate (0.15%) | 0.5 |
| 6% manganese naphthenate (0.015%) | 0.5 |
| 24% lead naphthenate (0.6%) | 5.0 |
| 4% calcium naphthenate (0.1%) | 6.0 |

## Utility Sealer/Finish

| | |
|---|---|
| "Keltrol" 1074 | 410.0 |
| Mineral spirits | 260.0 |
| VM&P naphtha | 45.0 |
| 6% cobalt naphthenate (0.03%) | 1.5 |

To decrease flow, 2-5% colloidal silica is added.

## Water-Repellent Paint

"Dow Corning" 772 water-repellent is an additive for solution concentrations as low as 0.1% based on the weight of solid material. When used as a water-repellent additive in latex paint, for instance, 1 lb, in 300 lb of paint will give a concentration of 0.1% siliconate. Optimum water repellency is usually developed with 0.1 to 1.0% concentration.

## Slick Finish

A mar-resistant surface or "slickness' can be imparted to a finish coat by incorporating about 1 oz. to the gallon of paraffin wax, petrolatum, or similar material. The wax is not compatible with the film; therefore, it exudes to the surface and gives the "slick" feel. It cannot be used in air-drying finishes because it retards oxidation tremendously. Also, it should not be used in baking undercoats because it detracts from the adhesion of subsequent coats.

The wax can be dissolved in a suitable solvent at concentrations of about 10%, and up to 10% of wax solution added for a concentration of 1% wax in the formulation. Caution should be observed, for too high a concentration causes some haze and tends to reduce the gloss as well as to produce an oily appearance on the surface; 1% of wax may be abnormally high.

---

## Epoxy Tar Paint

| | |
|---|---|
| Base component: | |
| Pigments | |
| "Asbestine" 3X | 252.1 |
| Microcrystalline silicate | 5.2 |
| Vehicle | |
| Coal tar pitch | 367.5 |
| "Epon" 834X-90 | 245.0 |
| Xylene | 65.4 |
| Secondary butyl alcohol | 65.4 |
| Curing-agent component: | |
| Diethylenetriamine | 31.3 |
| Secondary butyl alcohol | 31.3 |

The base component is prepared by heating the pitch to 120°F. The "Epon" is added together with the solvents. The two pigments are then blended in the mixing blender at low speed for 5 min. The two components are stored separately.

---

## FLAT EPOXY COATING
(Photochemically Nonreactive Solvent System)

### Epoxy Component

| | |
|---|---|
| "Genepoxy" 451CS60 | 356 |

### Coreactant Component

| | |
|---|---|
| "Diad" 30 | 127 |
| "Versamid" 230 | — |
| "Versamid" 230TP75 | 63 |
| "Tipure" R-902 | 570 |
| "Abestine" 325 | 580 |
| "Celite" 281 | 290 |

| | |
|---|---|
| Xylene | 18 |
| "Solvesso" 100 | — |
| "Solvesso" 150 | 24 |
| "Cellosolve" | 272 |
| Butyl "Cellosolve" | 60 |
| Mineral spirits | 190 |

**Clear Epoxy Coating**

| | |
|---|---|
| "Epon"-834 | 100 |
| Solvent | 40 |
| POPDA 230 | 30 |
| SR-82 | 1 |

**Curing Conditions:** Coated on steel test panel with 3-mil applicator blade; bake 45 min at 125°C.

**Epoxy Ester for Can Coatings**

| | |
|---|---|
| "ERL"-2773 | 64.750 |
| Disphenol A | 35.250 |
| Acids, dehydrated castor fatty | 7.880 |
| 1.4% Lithium naphthenate | 0.155 |
| Xylene | 4.300 |
| Acids, dehydrated castor fatty | 58.820 |
| Triphenyl phosphite | 0.330 |
| "Solvesso" 150 | 136.000 |
| Diacetone alcohol | 24.000 |

Charge "ERL"-2773, bisphenol A, first portion of DCO acids, lithium naphthenate, and xylene to still fitted with agitator, condenser, water trap and inert gas inlet. Blanket reactants with inert gas and heat rapidly to 390°F and allow exothermic reaction to occur. (A rise of 85-100°F will occur in approximately 5 min. Refluxing of the added xylene will control this exotherm). Cool to 460°F and hold at this temperature 30 min. Add the second portion of the DCO acids and the triphenyl phosphite and heat rapidly to 480°F. (It may be necessary to drain off part of the xylene originally added to reach 480°F.) Hold at 480°F with vigorous xylene reflux to Z viscosity at 50% n.v. in 85/15 "Solvesso" 150/diacetone alcohol (about 4-4 hr). Thin to 50% n.v. with same solvent blend.

**Epoxy Emulsion White Enamel**

| | |
|---|---|
| Emulsion (51.5% "Genepoxy" M195) | 330 |
| "TI-Pure" R-902 | 255 |
| Water | 240 |
| "Versamid" 265-WR70 | 210 |

**EPOXY ENAMEL FOR BRUSH APPLICATION**

(Photochemically Nonreactive Solvent System)

**Epoxy Component**

| | |
|---|---|
| "Genepoxy" 451CS60 | 677 |

| | |
|---|---|
| Diacetone alcohol | 88 |
| Butyl "Cellosolve" | 98 |

**Clear Varnish Component**

| | |
|---|---|
| "Versamid" 100P75 | 354 |
| "Diad" 30 | 95 |
| Mineral spirits | 100 |
| "Cellosolve" | 181 |
| "Solvesso" 150 | 24 |

**White Enamel Dispersion**

| | |
|---|---|
| "Versamid" 230TP75 | 190 |
| "Diad" 30 | 190 |
| "Tipure" R-902 | 460 |
| "Sparmite" | 230 |
| "Thixatrol" St. | 18 |
| Butyl "Cellosolve" | 19 |
| Mineral spirits | 161 |
| "Solvesso" 150 | 10 |

## ZINC-RICH EPOXY PRIMER, 80% ZINC

(Photochemically Nonreactive Solvent System)

**Ball Mill Base**

| | |
|---|---|
| Zinc dust No. 22 | 640.0 |
| "Bentone" 27 | 6.5 |
| Methanol | 3.0 |
| "Genepoxy" 525T75 | 139.0 |
| Butanol | 44.0 |
| 2-Nitropropane | 80.0 |
| VM and P (8% max. aromatic) | 43.0 |

**"Versamid" Resin Hardener**

| | |
|---|---|
| "Versamid" 115P80 | 70.0 |
| VM and P (8% max. aromatic) | 7.0 |

## PRIMER COATING, EPOXY POLYAMIDE, (CHEMICAL-AND SOLVENT-RESISTANT)

(Photochemically Nonreactive Solvent System)

**Pigmented Epoxy (Component I)**

| | |
|---|---|
| Strontium chromate | 12.0 |
| Titanium dioxide (Type III, Class A, TT-P-442) | 2.3 |
| Magnesium silicate, superfine | 5.5 |
| Diatomacious silica (MIL-S-1591) | 2.9 |
| "Genepoxy" 451T75 | 18.2 |
| Methyl isobutyl ketone (TT-M-268) | 2.0 |
| Xylene | 3.0 |
| 2-Nitropropane | 13.5 |
| Isopropanol | 2.7 |

**"Versamid" Resin**

| | |
|---|---|
| "Versamid" 115P80 | 9.4 |
| Isopropyl alcohol (GR.A, TT-I-735) | 9.0 |
| Butyl alcohol, normal (TT-B-846) | 7.0 |
| VM and P (8% max. aromatic) | 12.5 |

**Epoxy Emulsion Primer**

| | |
|---|---|
| 1. Emulsion (51.5% "Genepoxy" M195) | 306 |
| Water | 221 |
| Lead silico-chromate M50 | 290 |
| Talc | 75 |
| "Mapico" 387 (Pigment) | 140 |
| 2. "Versamid" 265-WR70 Pigment | |

**LEAD SILICO CHROMATE EPOXY PRIMER**

(Photochemically Nonreactive Solvent System)

| | |
|---|---|
| Lead silico chromate | 580.0 |
| "Bentone" 27 | 4.0 |
| "Versamid" 115P80 | 127.0 |
| Butanol | 17.0 |
| Mineral spirits | 81.0 |
| Toluene | 11.5 |

**Epoxy Component**

| | |
|---|---|
| "Genepoxy" 451CS60 | 316.0 |
| "Solvesso" 150 | 15.0 |
| Butanol | 11.0 |
| 2-Nitropropane | 40.0 |
| Toluene | 26.0 |
| Butyl cellosolve | 18.0 |

**Metal Primers**

| | No. 1 | No. 2 | No. 3 |
|---|---|---|---|
| Chlorinated polyolefin 310-6 resin | 10.0 | 9.85 | 8.55 |
| "Elvax" 40 wax | — | 5.01 | 4.35 |
| "Aroclor" 1260 plasticizer | — | 1.50 | 1.30 |
| "Aroclor" 1254 plasticizer | 5.6 | — | — |
| "Aroclor" 5460 resin | 2.7 | — | — |
| Stabilizer A-5 | 0.2 | 0.20 | 0.26 |
| Epichlorohydrin | 0.1 | 0.10 | 0.08 |
| Red lead pigment | 21.7 | 19.30 | 25.20 |
| "Bentone" 38 gelling agent | 1.3 | 1.08 | 1.40 |
| Xylene | 47.8 | 53.60 | 46.54 |
| Mica | 3.1 | 2.69 | 3.42 |
| Talc | 7.5 | 6.67 | 9.00 |

**Automotive Primer**

| | |
|---|---|
| Red iron oxide | 100.0 |
| China clay | 100.0 |
| Barytes | 200.0 |
| "Kesol" 1209 | 478.0 |
| 24% lead naphthenate (1.2%) | 16.8 |

| | |
|---|---|
| 6% cobalt naphtenate (0.03%) | 1.6 |
| 6% manganese naphthenate (0.03%) | 1.6 |
| "Activ"-8 (0.08%) | 0.3 |
| Water | 498.0 |

Bake 20 to 30 min at 275 to 325°F.

---

**Red Oxide Baking Primer**

| | |
|---|---|
| Charge pebble mill: | |
| "Stepanyl" 324-57E | 163.0 |
| Red iron oxide | 77.5 |
| Super microfine barytes | 170.0 |
| Aluminum silicate | 56.5 |
| Basic lead silicochromate | 56.5 |
| Deionized water | 92.5 |
| Pine oil | 66.2 |
| Grind 18 hr, led down: | |
| 6% Mn drier | 1.0 |
| "Stepanyl" 324-57E | 155.0 |
| Deionized water | 154.5 |
| n-Hexyl alcohol | 5.2 |
| 10% solution, water-soluble silicone | 2.1 |
| Water-soluble melamine (80% n.v.) | 59.5 |
| Antifoaming agent | 0.5 |

---

**Air-Drying Red Oxide Primer**

| | |
|---|---|
| Charge pebble mill: | |
| "Stepanyl" 324-57 | 140.0 |
| Red iron oxide (70% n.v.) | 90.0 |
| Barytes | 150.0 |
| Magnesium silicate | 50.0 |
| Basic lead silicochromate | 50.0 |
| Deionized water | 100.0 |
| Grind 18 hr, let down: | |
| "Stepanyl" 324-57 | 360.0 |
| 6% Mn drier | 1.3 |
| 24% Pb drier | 0.2 |
| Deionized water | 108.3 |

### Air-Drying Black Iron Oxide Primer

| Charge pebble mill: | |
|---|---|
| "Stepanyl" 324-57 | 114.0 |
| Rutile $TiO_2$ | 25.4 |
| Black iron oxide | 84.5 |
| Basic lead silicachromate | 67.5 |
| Barytes | 84.5 |
| Magnesium silicate | 42.2 |
| Ben-A-Gel'E | 4.2 |
| Odorless mineral splrits | 4.2 |
| Deionized water | 170.0 |
| Grind 18 hr, led down: | |
| 6% Mn drier | 0.9 |
| 24% Pb drier | 0.2 |
| "Stepanyl" 324-57 | 240.0 |
| Deionized $H_2O$ | 144.3 |
| n-Butanol | 18.1 |

### Salt-Spray Resistant Primer Additive

The addition of 1-2% "Aerosil" R-972 (based on pigments) greatly improves salt spray resistance of primers and other coating systems.

### Fiberglass Ceiling Tile Coating

| | |
|---|---|
| Water | 400.0 |
| R & R-551 | 5.0 |
| Butyl "Cellosolve" acetate | 8.0 |
| "Titanox" AMO | 200.0 |
| "Celite" 281 | 115.0 |
| No. 1 white | 190.0 |
| "Polyox" thickener WRS 301 1.5% solution | 45.0 |
| 9307 "Wallpol" | 190.0 |
| "Surfactol" 318 | 3.0 |

For spray application, 30 lb pressure, little atomizing pressure.

### Japan

| | |
|---|---|
| Cotton seed pitch | 25 |
| "Barber" gilsonite selects | 25 |
| Mineral spirits (35 KB) | 50 |

### Fireproof Paint

| | |
|---|---|
| Potassium silicate (35%) solution) | 41.5 |
| Aluminum orthophosphate | 0.3 |
| Water | 16.2 |
| Titanium dioxide | 24.9 |
| Aluminum hydroxide | 16.6 |
| Boric acid | 0.5 |

**Nonflammable Paint for Ship Interiors**

| | |
|---|---|
| "Parlon" S10 | 6.57 |
| Long oil soybean alkyd (60% in petroleum solvent)* | 21.08 |
| Titanium calcium pigment | 17.28 |
| Titanium dioxide | 14.34 |
| Antimony oxide | 8.86 |
| Yellow iron oxide | 0.59 |
| Zinc yellow | 0.93 |
| Lampblack | 0.08 |
| Xylene | 23.78 |
| Petroleum spirits | 6.33 |
| Coblat naphthenate | 0.13 |
| Epichlorohydrin | 0.03 |
| Antisag agent | as required |

---

**Nonskid Aircraft Carrier Flight Deck Coating**

| | |
|---|---|
| "Spenkel" M80-50CX | 48.00 |
| Isoamyl acetate | 6.00 |
| Xylene | 6.00 |
| Carbon black | 0.10 |
| Magnesium silicate, 325-mesh | 20.00 |
| Asbestos floats | 10.00 |
| G2-18-25 ground glass, 0.0394 to 0.0289-in. particles | 90.00 |
| G3-25-40 ground glass, 0.0280 to 0.0165-in. particles | — |

Mixing procedure consists of adding the isoamyl acetate to the xylene and then adding the mixture to the "Spenkel" M80-50CX. The ground glass particles are stirred into the diluted resin to insure thorough wetting of the glass particles. Finally, the remaining ingredients, the asbestine, asbestos floats and carbon black are blended together, added to the diluted resin-glass particle component and blended to a homogeneous mix.

---

**Marine Varnish**

| | |
|---|---|
| "Spenkel" F77-60MS | 650.00 |
| UV-absorber "Cyasorb-" UV-24 | 39.00 |
| Mineral spirits | 100.00 |
| 6% cobalt naphthenate (0.01%) | 0.65 |

| | |
|---|---|
| "Exkin" No. 2 (0.2%) | 0.80 |

**Ship-Bottom Paints**

No. 1

**Anticorrosion Primer**

| | |
|---|---|
| "Parlon" S20 | 2.4 |
| Coumarone indene resin, hard | 16.5 |
| Coumarone indene resin, soft | 4.2 |
| Coal tar | 4.6 |
| Zinc oxide | 14.0 |
| Zinc chromate | 9.5 |
| Mica | 3.7 |
| Indian red | 1.2 |
| Magnesium silicate | 15.2 |
| Aluminum stearate | 0.8 |
| Hi-Flash naphtha | 17.2 |
| Mineral spirits | 10.7 |

No. 2

**Antifouling Topcoat**

| | |
|---|---|
| "Parlon" S20 | 1.4 |
| Rosin WW (or WG or N) | 21.1 |
| "Hercolyn" D | 10.6 |
| Cuprous oxide | 42.2 |
| Diatomaceous silica | 7.1 |
| Hi-Flash naphtha | 17.6 |

No. 3

**Antifouling Topcoat**

| | |
|---|---|
| "Parlon" S125 | 2.50 |
| "Duraplex" C-45LV | 20.00 |
| Dibutyl phthalate | 0.30 |
| Titanium dioxide, rutile | 21.65 |
| Zinc oxide | 8.55 |
| Mercuric chloride | 17.60 |
| Carbon black | 0.15 |
| Copper resinate (8% copper) | 8.55 |
| Xylene | 20.30 |
| Lead naphthenate, 24% | 0.19 |
| Cobalt naphthenate, 6% | 0.08 |
| Phenoxy-propylene oxide | 0.13 |

## COATING, EPOXY-POLYAMIDE, CHEMICAL- AND SOLVENT-RESISTANT FOR WEAPONS SYSTEMS

(Photochemically Nonreactive Solvent System)

**Pigmented Epoxy (Component I)**

| | |
|---|---|
| Titanium dioxide (Type III, TT-T-425) | 30.8 |
| Silicone resin solution (60%, anticratering agent) | 0.3 |
| "Genepoxy" 451T75 | 26.6 |
| 2-Nitropropane | 8.2 |

**"Versamid" Resin**

| | |
|---|---|
| "Versamid" 115P80 | 14.4 |
| Butyl alcohol (TT-B-846) | 5.2 |

| | |
|---|---|
| Isopropyl alcohol | 6.0 |
| VM and P (8% max. aromatic) | 8.5 |

**Coil Coating Enamel**

| | |
|---|---|
| Rutile titanium dioxide | 225 |
| "Kelsol" 1688 | 320 |
| Melamine resin | 30 |
| Water | 465 |

If lower bake temperatures desired, refer to KELSOL 1688.

Add 3% tributyl phosphate on total enamel if bubbling is encountered.

Spraying at this consistency, blister-free film can be obtained with no flashoff before baking. If reduced further, 5-10 min flashoff necessary at film thicknesses greater than 1 mil dry.

**Nondrip Fire Retardant Polyethylene for Wire Coating**

| | |
|---|---|
| Polyethylene | 61.8 |
| "CHLOROWAX" 70 | 18.5 |
| Antimony trioxide | 18.5 |
| "Tribase" E | 1.0 |
| "Akroflex" C | 0.2 |

**PVC Dip Coating for Automotive Parts**

| | |
|---|---|
| "Exon" 654 | 100.00 |
| Ba-Cd-Zn stabilizer | 3.00 |
| Expoxidized soybean oil | 3.00 |
| Plasticizer 466 | 70.00 |
| "Bentone" 27 | 0.33 |
| Carbon black | 0.25 |

**Fuel-Tank Lining (Aircraft)**

| | |
|---|---|
| "Spenkel" M80-50CX | 700 |
| Xylol | 140 |

**Sound Deadening Acoustic Coating**

| | |
|---|---|
| Petroleum asphalt, 85-100 penetration vacuum reduced | 10.0 |
| "Barber" gilsonite VB | 2.5 |
| Petroleum naphtha | 12.5 |
| Asbestos 7R | 2.1 |
| Kaolin clay | 6.0 |
| Wetting agent | 0.2 |
| Ground limestone | 66.7 |

Apply by airless spray and dry by heating up to 450°F.

**Leather Finish**

| | |
|---|---|
| Pigment | 5.0 |
| U-3204, "Ubatol" 40% | 1 5 |
| U-3215, "Ubatol" 40% | 1.5 |
| Dye solution | 4.0 |
| Water | 4.0 |

**Interior Flat Tint Base**

| | |
|---|---|
| 3% "Methocel" Solution (65HG, 4000DG) | 100.0 |
| Polyglycol P-1200 | 2.0 |
| Water | 100.0 |
| PMA-18 | 1.0 |
| Surfactol 365 | 3.0 |
| "Aerosol" OT-B | 2.5 |
| "Titanox" RA-50 | 200.0 |
| Mica 325 mesh | 50.0 |

| | |
|---|---|
| "Celite" 281 | 25.0 |
| "Camel Carb" | 150.0 |
| "Foamicide" 581-B | 0.8 |
| 3% Methocel" solution (65HG, 4000DG) | 50.0 |
| "Carbitol" acetate | 20.0 |
| Water | 165.0 |
| "Everflex" BG | 275.0 |

---

**Pigment Dispersion**

No. 1 (Sudan Orange)

Charge to a ball mill:

| | |
|---|---|
| Sudan orange RA | 25.0 |
| "Isopar" H | 22.5 |
| "GANEX" V-220 | 2.5 |

Mill overnight

No. 2 (Carbon Black)

Dissolve "GANEX" V-220 in toluene. Add carbon black to solution with agitation.

Result: Approximately 30% dispersion of a free-flowing paste of carbon black.

---

**Paint Stripper (Caustic)**

| | |
|---|---|
| "Triton" QS-44 | 1.5 |
| Sodium hydroxide | 15.0 |
| Water | 83.5 |

Thoroughly mix the "Triton" QS-44 in the water. Then slowly add the sodium hydroxide with stirring.

---

**White Vinyl Strip Coat Lacquer**

| | |
|---|---|
| Methyl ethyl ketone | 132 |
| Xylol | 120 |
| Methyl isobutyl ketone | 21 |
| "Monoplex" S-38 | 28 |
| "Geon" 427 | 75 |
| "RBH" 5401 white paste | 7 |
| DC-200 silicone fluid | 1/8 |

Add the "Geon" resin slowly to the solvent while agitating a high-speed mixer. Continue stirring until the resin is completely dissolved, then add the remaining ingredients.

This lacquer should be sprayed full body. It is a temporary protective coating for use on wood or metal.

---

**Metallized Plastic Lacquer**

| | |
|---|---|
| "EHEC"-Low | 8.50 |
| "Amsco"-Solv D | 41.25 |
| V. M. & P. naphtha | 41.25 |
| Butanol | 9.00 |

---

**Wood Floor Lacquer**

| | |
|---|---|
| EHEC-Low | 7.5 |
| "Pentalyn" G | 7.5 |
| Ceresin wax | 0.2 |
| Mineral spirits | 76.3 |
| n-Propanol | 8.5 |

---

**Lacquer Protection Compound**

| | |
|---|---|
| "Ruhrwax" K-25 | 12 |
| Paraffin wax | 6 |
| Mineral spirits | 82 |

---

**Automobile Undercoating**

| | |
|---|---|
| "Ruhrwax" K-25 | 10 |
| "Ruhrwax" K-6 | 5 |
| Paraffin wax | 5 |
| Mineral spirits | 20 |
| Benzine | 60 |

The "Ruhrwaxes" K-25 and K-6 as well as paraffin are dissolved at a temperature of 100-115°C in the mineral spirit. The low boiling mineral spirit is benzine then added cold with agitation. After the entire quantity of mineral spirit has been added, the temperature should still be high enough (about 80°C) leaving the solution clear. Following this, the compound is cooled as quickly as possible with agitation and is homogenized below a temperature of 30°C. It is passed twice through the homogenizer.

---

**Clear Nitrocellulose Lacquer**

| | (dry basis) |
|---|---|
| Nitrocellulose, RS, 1/2 sec. | 14.4 |
| "Santolite" MHP | 5.7 |
| Camphor | 2.9 |
| Dibutyl phthalate | 4.4 |
| Isopropyl alcohol, anhydrous | 10.1 |
| n-Butyl acetate | 29.2 |
| Ethyl acetate | 17.9 |
| Toluene | 15.4 |

---

**Water-Vaporproof Heat-Sealing Lacquers**

| | No. 1 | No. 2 | No. 3 |
|---|---|---|---|
| "Parlon" S125 | 58.5 | 58.5 | 58.5 |
| "Lewisol" 28 | 15.0 | 15.0 | 15.0 |
| Dibutyl phthalate | 23.0 | 23.0 | 23.0 |
| Ceresin wax, white | 3.5 | — | — |
| Beeswax, crude | — | 3.5 | — |
| Paraffin wax (M.P. 120°F.) | — | — | 3.5 |

---

## PHOTOCHEMICALLY NONREACTIVE THINNERS

**Spray Thinner**

| | |
|---|---|
| Xylene | 6 |
| Toluene | 11 |
| 2-Nitropropane | 30 |
| Isopropanol | 14 |
| VM and P naphtha | 25 |
| Butanol | 14 |

**Brush Thinner**

| | |
|---|---|
| MIBK | 10 |
| "Solvesso" 150 | 6 |
| Mineral spirits | 25 |
| "Cellosolve" | 29 |
| Butyl "Cellosolve" | 10 |
| "Cellosolve" acetate | 10 |
| Butanol | 10 |

## Lacquer Thinners

Lacquer thinners usually are blends. The relatively more expensive pure solvents frequently are extended by adding a calculated proportion of nonsolvent or diluent to reduce the cost. A wide variety of raw materials are available for compounding. Diluents include aliphatic and aromatic hydrocarbons, while pure solvents include volatile aliphatic alcohols, esters, and ketones, as well as glycol-derived etheralcohols. In addition, there are medium and high-boiling solvents which fall outside the above classification.

The principal difficulty in using diluents is that of balancing boiling characteristics and evaporation rates. In a well formulated thinner it is desirable to balance the total evaporation time of the diluents and the pure solvents. Also, it is desirable to balance the total evaporation time of the alcohols and the esters-ketones. By doing this, one ensures the composition of the final evaporating film. For more information on solvents and their characteristics, the reader is referred to one of the various books on solvents which are available.

To simplify the calculation of the composltion of a balanced lacquer thinner formulation, the evaporation rate of each ingredient can be expressed in terms of n-butyl acetate, but with reference to 100 parts by weight of the ingredient. This figure is referred to in the following table as the Evaporation Rate Unit. In this table are listed some common lacquer thinner ingredients, densities in lb./gal., boiling points in °C., evaporation rates, and evaporation rate units as defined.

---

### Physical Data on Lacquer Thinner Ingredients

| Name | lb/gal | b.p. | evap. rate | evap. rate units |
|---|---|---|---|---|
| **Diluents** | | | | |
| Benzene | 7.31 | 79.6 | 500 | 20.0 |
| Toluene | 7.22 | 111.0 | 195 | 51.0 |
| Xylene | 7.20 | 135-45 | 68 | 147.0 |
| **Alcohols, ether alcohols** | | | | |
| Methanol | 6.60 | 64.5 | 370 | 27.0 |
| Ethanol | 6.60 | 78.5 | 203 | 49.0 |
| Isopropanol | 6.58 | 82.5 | 205 | 49.0 |
| n-Butanol | 6.75 | 117.1 | 45 | 222.0 |
| sec-Butanol | 6.72 | 99.5 | 115 | 87.0 |
| "Cellosolve" | 7.75 | 135.0 | 40 | 250.0 |
| Butyl "Cellosolve" | 7.50 | 170.6 | 10 | 1000.0 |

| | | | | |
|---|---|---|---|---|
| **Esters** | | | | |
| Methyl acetate | 7.80 | 59-60 | 1040 | 9.6 |
| Ethyl acetate | 7.37 | 77.1 | 525 | 19.0 |
| Isopropyl acetate | 7.28 | 89.0 | 435 | 23.0 |
| n-Butyl acetate | 7.29 | 126.5 | 100 | 100.0 |
| sec-Amyl acetate | 7.18 | 130-8 | 87 | 115.0 |
| "Cellosolve" acetate | 8.10 | 156.2 | 24 | 417.0 |
| **Ketones** | | | | |
| Acetone | 6.60 | 56.1 | 720 | 14.0 |
| Methyl ethyl ketone | 6.72 | 79.6 | 465 | 21.5 |
| Methyl isobutyl ketone | 6.67 | 118.0 | 145 | 69.0 |
| Cyclohexanone | 7.87 | 155-7 | 25 | 400.0 |

The formulas shown in the following table illustrate the use of evaporation rate data to obtain a balanced lacquer thinner. Using the first formula as an example, it will be seen that there are 35 parts of benzene by weight. Thus, evaporation rate units are 20 × 0.35 or 7.00. In the case of toluene, we find 20 parts by weight. Evaporation rate units thus are 51 × 0.20, or 10.20. The evaporation rate units for isopropanol are calculated similarly. Thus, 49 × 0.07 gives 3.43. Evaporation rate units are calculated for other ingredients in the same manner.

With further reference to the first formula, it will be noted that diluents show a total of 17.20 evaporation rate units while alcohols-esters-ketones wind up with 17.24, which is in good agreement. Alcohols total 8.65 units while esters-ketones total 8.60. Again, this indicates a practical degree of balancing.

---

### Asphalt Varnish

| | |
|---|---|
| China-wood oil | 24.8 |
| "Barber" gilsonite selects | 24.8 |
| Drier | 0.4 |
| Mineral spirits (35 KB) | 50.0 |

Other drying oils may be used depending upon requirements and cost limitations. The oil and Gilsonite are cooked together applying approximately the same procedures and temperatures as normally used for the individual oil in other similar formulations. Manufacturers in some cases merely blend Gilsonite solution with the oil without going through the cooking process.

## Lacquer Thinner Formulas

| Ingredients | % | Evap. Units | % | Evap. Units | % | Evap. Units | % | Evap. Units | % | Evap. Units |
|---|---|---|---|---|---|---|---|---|---|---|
| DILUENTS | | | | | | | | | | |
| Benzene | 35 | 7.00 | 30 | 6.00 | 25 | 5.00 | 20 | 4.00 | | |
| Toluene | 20 | 10.20 | 20 | 10.25 | 20 | 10.25 | 20 | 10.25 | 45 | 23.10 |
| Xylene | | | 5 | 7.35 | 10 | 14.70 | 15 | 22.10 | 10 | 14.70 |
| Evap. Units Total | | 17.20 | | 23.60 | | 29.95 | | 36.35 | | 37.80 |
| ALCOHOLS-Eth. Alc. | | | | | | | | | | |
| Methanol | | | | | 7 | 1.89 | | | | |
| Ethanol | | | | | | | 10 | 4.93 | 2 | 0.99 |
| Isopropanol | 7 | 3.43 | 3 | 1.45 | | | | | | |
| n-Butanol | | | | | 6 | 13.35 | | | 6 | 13.35 |
| sec-Butanol | 6 | 5.22 | 12 | 10.45 | | | 1 | 0.87 | 5 | 4.34 |
| "Cellosolve" | | | | | | | 5 | 12.50 | | |
| | | (8.65) | | | | | | | | |

| Ingredients | % | Evap. Units | % | Evap. Units | % | Evap. Units | % | Evap. Units | % | Evap. Units |
|---|---|---|---|---|---|---|---|---|---|---|
| ESTERS | | | | | | | | | | |
| Methyl Acetate | | | | | | | 6 | 0.58 | | |
| Ethyl Acetate | | | | | 15 | 2.86 | | | | |
| Isopropyl Acetate | 20 | 4.60 | 15 | 3.45 | | | | | 11 | 2.58 |
| n-Butyl Acetate | | | | | | | 9 | 9.00 | | |
| sec-Amyl Acetate | | | | | 6 | 6.90 | | | 8 | 9.25 |
| "Cellosolve" Acetate | | | | | | | 1 | 4.16 | | |
| KETONES | | | | | | | | | | |
| Acetone | | | | | | | 3 | 0.42 | | |
| Methyl Ethyl Ketone | 9 | 1.93 | 4 | 0.86 | 5 | 1.07 | 6 | 1.29 | 4 | 0.86 |
| Methyl Isobutyl Ketone | 3 | 2.07<br>(8.60) | 11 | 7.58 | 6 | 4.14 | 4 | 2.76 | 9 | 6.20 |
| Totals | 100 | 17.24 | 100 | 23.79 | 100 | 30.21 | 100 | 36.51 | 100 | 37.57 |

### Colored Asphalt Paint

| | |
|---|---|
| 85 to 100 penetration steam-vacuum asphalt | 17.5 |
| Gilsonite selects | 17.5 |
| Mineral spirits (35 KB) | 40.0 |
| Metallic oxide pigment, green, red, brown, yellow | 25.0 |

The pigment is incorporated in ball mills or other dispersing equipment. Again the intensity depends upon the amount of pigment used. Percentages as low as 15% by weight do give an appreciable color but ordinarily approximately 25% is required.

---

### Driveway Sealer

| | |
|---|---|
| 85 to 100 penetration steam-vacuum asphalt | 24.0 |
| "Barber" gilsonite selects | 16.0 |
| Mineral spirits (35 KB) | 60.0 |

---

### Asphalt Products, Improving

"Paricin" 285 at a 1% level in asphalt raises the softening point and lowers the viscosity of the melt without increasing brittleness. It also improves the acid, water and salt spray resistance of the asphalt.

---

### Asphalt Paint

No. 1

| | |
|---|---|
| 85 to 100-penetration steam-vacuum asphalt | 25.0 |
| "Barber" gilsonite selects | 25.0 |
| Mineral spirits (35 KB) | 50.0 |

No. 2 (Pigmented)

| | |
|---|---|
| 85 to 100-penetration steam-vacuum asphalt | 23.5 |
| "Barber" gilsonite selects | 23.5 |
| Mineral spirits (35 KB) | 47.0 |
| Carbon black (pigment grade) | 6.0 |

Disperse in ball mill.

---

### Asphalt Aluminum Paint

| | |
|---|---|
| 85 to 100-penetration steam-vacuum asphalt | 17.5 |
| "Barber" gilsonite selects | 17.5 |
| Mineral spirits | 35.0 |
| Xylol | 5.0 |
| Aluminum paste | 25.0 |

## Spackle

| | | |
|---|---|---|
| Powdered Base | | |
| U.S. gypsum plaster | 83.95 | |
| Paragon clay | 12.61 | |
| Dry glue | 4.26 | |
| Zinc sulphate | 0.18 | |
| | 100.00 | 76.00 |
| Regular silica | | 6.50 |
| "Pyrax" A | | 17.00 |
| Terra alba No. 1 | | 0.50 |
| | | 100.00 |

Mix and grind on Buhr stone mill.

### Washable Calcimine

| | |
|---|---|
| Casein | 8.00 |
| Lime | 8.00 |
| Pine oil | 1.00 |
| "Cryptone" No. 19 | 10.00 |
| Micronized bentonite | 1.00 |
| "Pyrax" A | 15.00 |
| Royal blue | 0.02 |
| Paragon clay | 12.00 |
| Factory whiting | 10.00 |
| No. 26 whiting | 34.98 |

### Extra Fine Quality Calcimine

| | |
|---|---|
| Factory whiting | 75.00 |
| Paragon clay | 12.80 |
| U.S. gypsum plaster | 7.75 |
| Dry glue | 4.27 |
| Zinc sulphate | 0.18 |

Grind on a Buhr stone mill.

### Casein, Gum, and Linseed Oil Emulsion

| | |
|---|---|
| Heavy casein solution* | 130 |
| Water | 50 |
| Ester gum solution (Gum—60.92%, min. spir. 39.08%) | 144 |
| Kerosene | 22 |
| Linseed oil | 15 |
| Pine oil | 4 |
| "Preventol" solution (15% in water) | 8 |

### Casein Water Paints

No. 1

Exterior White

| | |
|---|---|
| Heavy casein solution* | 370 |
| 10% Gum karaya solution | 8 |
| Water | 60 |
| Lithopone | 500 |
| "Dicalite" L | 90 |
| Barytes | 45 |

*Heavy Casein Solution

| | lbs. |
|---|---|
| Casein, oiled | 75.00 |
| Water | 300.00 |
| Sodium fluoride | 8.40 |
| Borax | 8.25 |
| "Preventol" solution 15% in water | 1.00 |

No. 2

Washable White

| | |
|---|---|
| Casein, oiled | 10.50 |

| | |
|---|---|
| Pine oil | 1.00 |
| GC sodium silicate | 4.00 |
| Univ. spray lime | 1.00 |
| "Cryptone" No. 19 | 35.00 |
| No. 232 Talc | 11.00 |
| "Pyrax" TES (1) | 15.00 |
| No. 1291 Clay | 8.00 |
| Lion clay powder | 9.30 |
| "Pyrax TES" (2) | 5.00 |
| "Dowicide" A | 0.20 |

The pine oil and talc are mixed together. The silicate and "Pyrax" TES (2) are mixed and passed through the Quaker City mill. The casein, No. 1291 clay, lion clay, "Pyrax" TES (1) and "Dowicide" A are weighed out. Mix all and grind through a Buhr stone mill.

No. 3

Improved Washable White

| | |
|---|---|
| Casein, oiled | 10.50 |
| "Dowcide" A | 0.20 |
| Hydrated lime | 5.00 |
| No. 26 Whiting | 19.90 |
| Paragon clay | 5.00 |
| "Pyrax" TES | 5.00 |
| Pine oil | 1.00 |
| "Unilith" 22N | 33.40 |

The pine oil and the paragon clay are mixed together before the other materials are incorporated.

**Casein White Paste Paints**

| | No. 4 | No. 5 |
|---|---|---|
| Heavy casein stable paste* | 400.00 | 42 |
| "Cryptone" No. 19 | 22.10 | — |
| L. O. Beckton white | 5.00 | 38 |
| No. 232 Talc | 8.70 | 5 |
| No. 1291 English china clay | 8.70 | 5 |
| "Pyrax" A | 15.00 | 10 |
| Glycerine | 0.50 | — |
| | 100.00 | 100 |

The powders are mixed in a Baker-Perkins mixer and then sufficient vehicle or paste added to make a stiff paste. This is then ground for some 25 min before the remainder of the vehicle or paste is added. From the mixer the paste is passed through either a Charlotte or Eppenbach colloid mill and finally out through a two-way spigot into awaiting cans for packaging.

*Heavy Casein Stable Paste

| | |
|---|---|
| Casein, oiled | 17.50 |
| Diatomacious earth, purified | 1.33 |
| Conc. ammonium hydroxide | 1.10 |
| Ammonium hydrofluoride | 0.83 |
| Water | 72.74 |
| Butyl "Cellosolve" | 2.50 |
| Dibutyl phthalate | 1.50 |
| Hexalin | 0.50 |
| Glycerine | 0.50 |
| Pine oil | 1.50 |

Chapter V

# COSMETICS AND DRUGS

## BATH PREPARATIONS

### Bath Oil

No. 1

| | |
|---|---|
| Isopropyl myristate | 74.0 |
| "Robane" | 10.0 |
| Lanolin oil | 5.0 |
| Mineral oil and perfume q.s. | 100.0 |

No. 2

| | |
|---|---|
| Lantrol | 25.0 |
| Isopropyl palmitate | 25.0 |
| Perfume | 50.0 |

With agitation mix the perfume in with the oils to get complete solution. Filter, polish, and fill.

No. 3

| | |
|---|---|
| "Lantrol" | 5.0 |
| Polyethylene glycol 400 dioleate | 2.5 |
| Isopropyl myristate | 9.5 |
| Mineral oil—70 vis. | 79.5 |
| D & C Green No. 6 (0.25% in isopropy myristate) | 0.5 |
| Perfume | 3.0 |

Bath Cube

| | |
|---|---|
| Talc | 5.0 |
| Starch | 8.0 |
| Borax | 2.0 |
| Soda ash | 24.0 |
| Sodium bicarbonate | 24.0 |
| Perfumer and color to suit | |

No. 4

| | |
|---|---|
| "Solar" 25 | 20.0 |
| "Blandol" | 60.0 |
| "Brij" 93 | 13.0 |
| "Ninol" AA-62 extra | |
| "Ninol" AA-62 extra | 2.0 |
| Perfume | 5.0 |

### Bubble Baths

| | No. 1 | No. 2 | No. 3 |
|---|---|---|---|
| 40% Triethanolamine lauryl sulfate | 40.0 | 40.0 | 40.0 |
| "Aromox" C/12-W | 5.0-10.0 | | |

| | | | |
|---|---|---|---|
| "Aromox" DMC-W | | 5.0-10.0 | |
| "Aromox" DMMC-W | | | 5.0-10.0 |
| Water | to 100 | to 100 | to 100 |
| Perfume and preservatives | Q.S. | Q.S. | Q.S. |

These formulas are prepared by mixing all ingredients together while warming gently.

No. 4

| | |
|---|---|
| "Plurafac" A-38 or C-17 | 20 |
| $Na_2CO_3 \cdot NaHCO_3 \cdot 2H_2O$ | 40 |
| $(NaPo_3)_6$ | 15 |
| $Na_2SO_4$ | 15 |
| Corn Starch | 10 |
| Perfume | trace |

---

**Shampoo**

No. 1

(Sulfonated Oil)

| | |
|---|---|
| "Sulfonated" castor oil | 25 |
| "Emcol" 5100 | 75 |

No. 2

(Sulfonated Oil-Soapless)

| | |
|---|---|
| Sulfonated castor oil (100% basis) | 125 |
| Sulfonated olive oil (100% basis) | 125 |
| Sulfonated coconut oil (100% basis) | 125 |
| Polyethylene glycol 1000 | 10 |
| Morpholine | 10 |
| Water | 600 |
| Perfume | |

Mix first three ingredients, heat to 104°F and stir in the morpholine and the polyethylene glycol 1000, followed by the water. Add perfume when cool.

| | No. 3 | No. 4 |
|---|---|---|
| "Triton" X-200 | 50.00 | 50.00 |
| "Sanderol" 200L | 5.00 | 4.00 |
| Isopropyl palmitate | 0.50 | — |
| "Triton" X-400 conc. | — | 0.50 |
| Methyl "Paraben" | 0.15 | 0.15 |

| | | |
|---|---|---|
| EDTA | 0.05 | 0.05 |
| Water | 44.30 | 45.30 |

The EDTA and methyl "Paraben" are dissolved in water by heating to 160°F. The "Triton" X-200 is then added with stirring followed by the isopropyl palmitate or "Triton" X-400 conc., and finally the "Sanderol" 200-L is stirred in. The mixture is allowed to cool to 120°F, at which point perfumes and coloring agents can be added. In both formulations the order of addition is important.

No. 5

This formulation has been found to be particularly effective for washing oily hair.

| | |
|---|---|
| "Triton" X-200 | 36 |
| Sodium lauryl sulfate | 32 |
| "Triton" X-15 | 2 |
| Water | 30 |

"Triton" X-15 is slowly stirred into the "Triton" X-200. The sodium lauryl sulfate and water are added in that order with thorough mixing to prevent separation of the shampoo on prolonged standing.

| | No. 6 | No. 7 | No. 8 | No. 9 | No. 10 |
|---|---|---|---|---|---|
| "Standapol" T | 40% | — | 20% | 30% | — |
| "Standapol" WAQ special | — | 50% | — | — | 10% |
| "Standapol" ES-2 | — | 15% | 20% | 10% | — |
| "Standapol" ES-40 concentrate | — | — | — | — | 15% |
| "Monamid" 150 LW-C | 4% | 5% | — | 3% | 2% |
| "Aromox" DMCD | — | — | 2% | — | — |
| "Standamol"-01 | — | — | 1% | 1% | — |
| Water, preservative, perfume, color | q.s. | q.s. | q.s. | q.s. | q.s. |

| | No. 11 | No. 12 (Liquid) | No. 13 (Cream) |
|---|---|---|---|
| Coconut-oil fatty acids | 42 lb. | 42.0 lb. | 21.0 lb. |
| Oleic acid | 56 | 56.0 | 14.0 |
| Triple-pressed-stearic acid | — | — | 13.5 |
| Propylene glycol | 55 | 55.0 | — |
| Monoethanolamine | — | 12.6 | 7.5 |
| Triethanolamine | 58 | 28.5 | 14.2 |

| | | | |
|---|---|---|---|
| 40% Formalin | — | — | 1.2 |
| Titanium dioxide | — | — | 0.2 |
| "Surfonic" N-40 | — | 10.0 | 3.0 |
| Hydroxyethyl cellulose | — | — | 35.0 |
| Distilled water | — | — | 125.0 |

To prepare formulas 11 and 12, mix the fatty acids, add the amines, and then the propylene glycol. Stir until a clear solution is obtained; then add the "Surfonic" N-40 and formalin, if desired. No heating is required. Dilute with water to any desired consistency. When the water is first added, the shampoo concentrate assumes a petrolatum-like consistency, but gradually changes to a clear, very slightly viscous solution of pale amber color. If the water solution is cloudy, stir in more amine, a little at a time, until it becomes clear.

A solution of 1 part of either formula with 3 parts by weight of water before use makes an excellent shampoo.

To prepare formula 13, melt the fatty acids together and adjust the temperature to 50 to 55°C. Add the "Surfonic" N-40 and the formalin. Heat the water to 60°C; add the amines and the hydroxyethyl cellulose. Adjust the temperature of the solution to about 50°C. Add the water solution to the melted fatty acids and stir constantly until a clear, viscou mixture is obtained and then at intervals until the temperature is about 35°C.

Add the perfume and mix thoroughly. Disperse the titanium dioxide in half of its weight of "Surfonic" N-40, grind it to a smooth paste, and then blend it into the cream. It may be necessary to mill the final product in order to obtain a complete dispersion of the titanium dioxide in the cream. Prepare a concentrated solution of a suitable dye. Add a small amount of this solution at a time, with thorough mixing, until the cream is of the desired tint.

Amine soaps will darken in color on standing, though this has no effect on their properties. A better color is maintained if the concentrated shampoo is diluted with water when first made than if it is stored and diluted at some later time. The addition of 0.5 to 1% borax or trisodium phosphate to the water, when diluting the shampoo, will prevent some of the discoloration. The addition of one half lb of formalin (40% formaldehyde solution) to each 200 lb of concentrated shampoo will prevent much of the discoloration during storage. The formalin also seems to decrease the soapy odor of the shampoo. Treated shampoo should be shelf-tested for a few weeks to ascertain whether the formalin will affect the perfume normally used in such products.

No. 14

Professional Shampoo Concentrate

| | |
|---|---|
| 60% sodium myristyl ether sulfate | 21 |
| 40% Triethanolamine lauryl sulfate | 20 |
| Lauric-myristic diethanolamide | 19 |
| Sodium chloride | 7 |
| Propylene glycol | 5 |
| Water, perfume, dye, preservative | 28 |

To be diluted, one pint to a gallon with water.

No. 15

Low Eye Irritation

| | | |
|---|---|---|
| "Miranol" $C_2M$ | 18.10 | I |
| "Sipon" LT6 | 23.00 | |
| Propylene glycol | 10.00 | |
| "Ethoxyol" 16 | 3.50 | II |
| "Lantrol" | 1.00 | |
| "Schercomid" SCO extra | 3.00 | |
| Water | 41.40 | III |

Heat I to 78°C. Heat II to 78°C slowly, with agitation. Add II to I. Heat III to 78°C. Add to I, II. Agitate, cool to 50°C. Perfume, agitate. Cool to 30°C.

No. 16

For Damaged Hair

| | |
|---|---|
| Phase I | |
| Triethanolamine lauryl sulfate (40%) | 30.0 |
| Deionized water | 46.0 |
| Phase II | |
| Deionized water | 5.0 |
| Propylene glycol | 5.0 |
| Triethanolamine | 1.7 |
| Oleic acid | 3.3 |
| Phase III | |
| "Ethoxyol" 16 | 3.0 |
| "Lantrol" | 1.0 |
| "Schercomid" SCO extra | 5.0 |

Phase I is heated to 75—80°C. Phase II is prepared by heating the first three ingredients to 75°C, then slowly adding the oleic acid with stirring. When clear this is added to Phase I. Phase III is prepared by heating the "Ethoxyol" 16 and "Lantrol" to 75—80°C, adding the diethanolamide (at room temperature) and reheating to 75—80°C. Phase III is then added to the previously combined Phases I and II, slowly and with stirring. If the final product is not clear when all of Part III is added, a few minutes additional heating at 75—80°C will make it clear.

No. 17

Antidandruff

| | |
|---|---|
| "Veegum" K | 3.0 |
| Water | 40.0 |
| "Vancide" 89RE | 2.0 |
| "Solulan" 98 | 1.0 |
| Cocoyl sarcosine | 4.0 |
| "Igepon" TC-42 (24% solids) | 50.0 |

Add the "Veegum" K to the water slowly, continually agitating until smooth. Combine B and until the "Vancide" 89RE is dispersed. Add II to C. Mix until uniform. Slowly add I to III.

Note: The final pH should be about 5.0.

No. 18

Emollient

| | |
|---|---|
| "Tegobetaine" C | 12.5 |
| Water | 48.1 |
| Tea lauryl sulfate (40%) | 18.3 |
| Na lauryl ether sulfate (28%) | 14.1 |
| Coco diethanolamide | 5.0 |
| "Tegester" 504-D | 2.0 |
| Citric acid | qs |

No. 19

Antiseptic

| | |
|---|---|
| "Tegobetaine" C | 37.50 |
| Water | 43.65 |
| Tea lauryl sulfate (40%) | 6.10 |
| Na lauryl ether sulfate (28%) | 4.70 |
| "Versene" 100 | 0.05 |
| Coco diethanolamide | 5.00 |
| Propylene glycol | 2.50 |
| Hexachlorophene | 0.50 |

No. 20

Baby Shampoo (No Sting)

| | |
|---|---|
| | I |
| "Miranol" $C_2M$ | 18.1 |
| "Sipon" LT6 | 19.2 |
| Water | 45.2 |
| Propylene glycol | 10.0 |
| | II |
| "Ethoxyol" 16 | 3.5 |
| "Lantrol" | 1.0 |
| "Schercomid" SCO Extra | 3.0 |

Heat I and II to 78°C. Add II to I, agitate. Cool 50°C. Perfume. Cool and agitate to 30°C.

**Cream Hair Rinse**

No. 1

| | |
|---|---|
| "Triton" X-400 | 7.5 |
| Cetyl alcohol | 0.3 |
| Potassium chloride | 0.8 |
| Water | 91.4 |

No. 2

| | |
|---|---|
| "Triton" X-400 Concentrate | 2.3 |
| Cetyl alcohol | 0.3 |
| Potassium chloride | 0.8 |
| Water | 96.6 |

The pH is 4.4.

This cream rinse should be diluted for use at the rate of 1 tablespoonful to 8 oz. of water.

Approximately half of the water is heated to 180°F. The cetyl alcohol is added and stirred slowly until molten. Slow speed agitation is used during the entire manufacturing process. Care should be taken to prevent the incorporation of air into the product.

The "Triton" X-400 is heated to 170°F and added to the water slowly. The temperature of the mixture is then reduced to 170°F and is agitated slowly for at least 30 min to ensure uniformity and maximum emulsion stability. The temperature is further reduced to 120°F.

The potassium chloride is dissolved in the balance of the water and heated to 120°F. This solution is then slowly added to the "Triton" X-400-cetyl alcohol emulsion. The heat source is removed, and the batch is agitated slowly for an additional hour, or until the mixture gels.

Coloring agents and perfumes solubilized in "Triton" X-100 may then be added.

No. 3

| | |
|---|---|
| "Triton" X-400 (25%) | 12.0 |
| Cetyl alcohol | 1.0 |
| Glycerol monostearate | 1.0 |
| Cosmetic polypeptides (55%) | 9.0 |
| "Hyamine" 3500 (50%) | 0.5 |
| Water | 76.5 |

No. 4

| | |
|---|---|
| "Triton" X-400 | 45.0 |
| "Triton" X-15 | 2.5 |
| Potassium chloride | 1.0 |
| Water | 51.5 |

Warm "Triton" X-400 to 180°F. Add 1/10 of water and the "Triton" X-15 to the hot "Triton" X-400 with agitation. Dissolve KCl in remaining water and add slowly with continuous agitation to insure a homogeneous mix. Dilute this formulation 1 part to 7 parts water before packaging.

No. 5

| | |
|---|---|
| "Triton" X-400. | 12.0 |
| Cetyl alcohol | 1.0 |
| Glycerol monostearate | 1.0 |
| Cosmetic polypeptides (55%) | 9.0 |
| Preservative | |

| | |
|---|---|
| ("Hyamine" 3500-50%) | 0.5 |
| Water | 76.5 |

No. 6

Pearly

| | |
|---|---|
| "Richamate" 3780 | 6.8 |
| PEG 400 distearate | 1.1 |
| Sodium sulfate | 0.5 |
| Formalin | 0.1 |
| Dye, perfume, and water q.s. | 100.0 |

Add the first four components to water and heat to 60°C with continuous mixing. Cool to 45°C and add dye and perfume.

No. 7

Softener

| | |
|---|---|
| "Adogen" 432-CG | 3.5 |
| Perfume | q.s. |
| Dye | q.s. |
| Deionized water | 100.0 |

No. 8

| | |
|---|---|
| "Adogen" 432 CG | 2.5 |
| Perfume | q.s. |
| "Adol" 52 | 1.0 |
| Dye | q.s. |
| Deionized water | q.s. to 100 |

This is prepared by heating the water to 80°C and adding the "Adogen" and the "Adol" to the water with agitation. The agitation is continued until the mass is cooled to room temperature.

## Skin Cleaners

No. 1

Creams

| | |
|---|---|
| Stearic Acid | 100.0 |
| Mineral oil | 500.0 |
| Lanolin | 70.0 |
| Terpineol | 1.0 |
| Triethanolamine | 12.5 |
| Propylene glycol | 50.0 |
| Water | 250.0 |
| Preservative | 10.0 |
| Perfume | |

Heat water to 158°F, adding triethanolamine. Heat first four ingredients and glycol at same temperature until melted, and stir water solution into them. Continue stirring until mixture cools to 122°F, then stir in preservative and perfume.

No. 2

| | |
|---|---|
| Stearic Acid | 4.0 |
| "Robane" | 28.0 |
| Glycerin | 1.0 |
| Triethanolamine | 1.0 |
| Preservative | 0.2 |
| Water q.s. | 100.0 |

No. 3

Phase A

| | |
|---|---|
| "Lantrol" | 5.0 |
| Castor oil, USP | 35.0 |
| "Arlacel" 60 (sorbitan monostearate) | 2.0 |
| Mineral oil heavy USP 355 vis. | 36.7 |
| Stearic acid, USP | 4.0 |
| "Tegosept" P | 0.2 |

Phase B

| | |
|---|---|
| Triethanolamine | 0.9 |
| $H_2O$ | 15.0 |
| "Tween 60" | 1.0 |
| "Tegosept" M | 0.2 |

Heat Part A to 78°C; heat Part B to 78°C. Add B to A. Cool with agitation to 50°C. Add perfume, cool with hand agitation to 25°C. Fill.

No. 4

| Textured | |
|---|---|
| "Veegum" | 1.75 |
| Water | 56.50 |
| Beeswax | 1.32 |
| Spermaceti | 1.32 |
| Light mineral oil | 17.40 |
| Sorbitan monopalmitate | 3.05 |
| Polysorbate 60 | 3.05 |
| Cetyl alcohol | 2.61 |
| "Nytal" 100 | 13.00 |
| Preservative | q.s. |

Add the "Veegum" to the water slowly, agitating continually until smooth. Heat A to 75°C. Heat B to 80°C. Add II to I and mix thoroughly. Add C to III and continue mixing until temperature reaches 40°C.

No. 5

| Glossy | |
|---|---|
| "Veegum" | 1.5 |
| Water | 54.0 |
| Propylene glycol | 4.0 |
| Glyceryl monostearate SE | 7.0 |
| "Pluronic" F 68 | 5.0 |
| Light mineral oil | 25.0 |
| "Lanacel" | 0.5 |
| "Trisolan" | 2.5 |
| "Vanseal" CS | 0.5 |
| Preservative | q.s. |

Add the "Veegum" to the water slowly, agitating continually until smooth. Add the propylene glycol and heat to 70°C. Heat B to 75°C. Add II to I and continue mixing until temperature reaches 40°C.

No. 6

| Emollient | |
|---|---|
| "Veegum" | 1.0 |
| Water | 50.7 |
| Cetyl alcohol | 0.3 |
| Stearyl alcohol | 0.3 |
| "Lanacet" | 1.0 |
| "Nimlesterol" D | 5.0 |
| Stearic acid | 4.4 |
| "Vanseal" Cs | 3.3 |
| "Pluronic" F 68 | 12.0 |
| "Igepon" AC78 | 20.0 |
| "Aromox" *K*/12W | 2.0 |
| Preservative | q.s. |

Add the "Veegum" to the water slowly, agitating continually until smooth. Heat A to 75°C. Heat B to 70°C, with slow mixing until uniform. Add I to II with slow agitation. Allow to cool to 40°C and package while warm.

---

**Liquid Cleaning Cream and Hand Lotion**

The high percentage of triethanolamine soap used in this liquid cleansing cream serves to emulsify completely the white mineral oil and lanolin, aids their penetration into the pores of the skin, and produces a cream that cleanses the skin yet is readily removed with water. Because of the high lanolin and propylene glycol content of this cream, it can be used as a hand lotion and all-purpose cream, as well as a cleansing cream. The lanolin is especially soothing to chapped or dried skin and, though the cream requires more massaging than some hand lotions to rub into the skin and eliminate stickiness, the softening and healing of the skin compensates for the extra time required for application of the cream.

| | | | |
|---|---|---|---|
| Stearic acid | 25 lb. | Triethanolamine | 9.5 lb. |
| White mineral oil | 57 | Propylene glycol | 75.0 |
| Anhydrous lanolin | 34 | Quince-seed mucilage | 19.0 |
| Terpineol | 1/3 | Water | 315.0 |

Melt the stearic acid in the mineral oil, add the lanolin and terpineol, and bring the temperature of the solution to 70°C. In a separate container, bring the solution of the triethanolamine in the water to 70°C. Add the hot oil mixture to the heated amine solution and stir vigorously until a good emulsion is formed. Add the quince-seed mucilage, made by adding 9¼ oz. of quince seed to 20 lb of water at 80°C, soaking overnight, and then straining through a cloth. A suitable preservative should be added to the quince-seed mucilage to prevent its molding.

Mix the perfume in the propylene glycol and stir this solution into the cream when it has cooled to about 50°C. The stirring should be fast enough to keep the cream mixed but not to aerate it. Continue stirring at low speed until the emulsion has cooled to room temperature. If the cream is allowed to cool without stirring, it will thicken on standing a few days.

The mineral oil can be replaced in its entirety or in part with a vegetable oil, such as olive or sweet-almond oil. The lanolin content can be decreased slightly where these oils replace some of the mineral oil.

Polyethylene glycol-1000 has been found to be an excellent thickening and stabilizing agent for liquid creams. It requires no preservative or special preparation. A dispersion of karaya gum or sodium alginate may also be used in place of the quince-seed mucilage. A dispersion of desirable consistency and with the slippery feel of the quince-seed mucilage can be prepared by stirring ½ lb of sodium alginate into 50 lb of hot water containing 1 lb of triethanolamine. The alginate is added slowly, with rapid stirring, until a smooth dispersion is obtained. A preservative should be added to the dispersion. Karaya gum can be dispersed in a similar manner, but the dispersion is thinner, so that less water should be used with this gum.

---

**Facial Wash Cream**

| | |
|---|---|
| Light mineral | 17.0 |
| Liquid base, (Croda) | 5.0 |
| "Polawax" | 5.0 |
| "Volop" 20 | 8.0 |
| Stearyl alcohol, 95% | 2.0 |
| "Tegamine" S-13 | 0.5 |
| "Tegamine" P-13 | 0.5 |
| Propylene glycol | 12.0 |
| Water, demineralized | 50.0 |
| Phosphoric acid to pH 5.0-5.5 | q.s. |

Add the oils to the water phase at about 75 to 80°C, begin cooling to R.T., and adjust to the acid side with the phosphoric acid.

---

**Cosmetic Cleaning Bar**

No. 1

| | |
|---|---|
| A | |
| "Vancide" 89RE | 1.0 |
| "Igepon" AC-78 (83% solids) | 39.0 |
| B | |
| Citric acid | 0.5 |
| "Igepal" DM-970 | 50.0 |
| Water | 9.5 |

Blend the "Vancide" 89RE in part of the "Igepon"; then add balance of "Igepon" and blend well. Heat B to 70-75°C. Add I to II with agitation, mix until uniform. Press into bar or cake.

Note: The final pH should be about 5.0.

No. 2

| | |
|---|---|
| A | |
| "Vancide" 89RE | 1.0 |
| "Veegum" F | 1.0 |
| "Igepon" AC-78 (83% solids) | 57.3 |
| B | |
| Cetyl alcohol | 2.0 |
| Glycerol monostearate, acid stable | 5.5 |
| Stearyl alcohol | 7.5 |
| "Modulan" | 3.0 |
| Polyethylene glycol 6000 | 13.0 |
| Citric acid | 0.7 |
| Water | 9.0 |

Blend the "Vancide" 89RE and "Veegum" F in part of "Igepon"; then add balance of "Igepon" and blend well. Heat B to 70-75°C. Add I to II with agitation until uniform. Press into bar or cake.

---

**Transparent Gel**

No. 1

| | |
|---|---|
| Phase I | |
| "Carbopol" 934 | 0.75 |
| "Sequestrene" $Na_4$ | 0.01 |
| Water | 79.45 |
| Phase II | |
| SDA 40 | 12.14 |
| Triethanolamine | 0.60 |
| Lantrol AWS | 2.00 |
| "Uvinul" 400 | 0.05 |
| "Carbowax" 400 | 5.00 |

Phase I passed through mill.

Eppenbach setting 20, to obtain Carbopol 934 solution. Agitate Phase II till solution of ingredients. Add Phase II to I with agitation. May be diluted to desired viscosity.

---

**Cleansing Lotion**

| | |
|---|---|
| Mineral | 125 |
| Polyethylene glycol 400 monostearate | 110 |
| Beeswax | 100 |
| Lanolin | 50 |
| Propylene glycol | 50 |
| Triethanolamine | 10 |
| Water | 540 |
| Preservative | 10 |
| Perfume | |

Heat first five ingredients together to 167°F until uniform. Add triethanolamine to water, heat to same temperature, and add to first mixture with vigorous tirring. When cooling to 104°F has occurred, add preservative and perfume.

---

**Vanishing Cream (With Lanolin)**

No. 1

| | |
|---|---|
| Stearic acid | 190 |
| Glycerine | 25 |
| Lanolin | 20 |
| Triethanolamine | 10 |
| Water | 800 |
| Preservative | 10 |
| Perfume | |

Melt together the stearic acid and lanolin and heat to 176°F. Add the triethanolamine and glycerin to the water, heat to 176°F, and pour into the first mixture with constant stirring for one hour. When cooling to 140°F has occurred, add preservative and when cool ng to 104°F has occurred, stir in perfume.

No. 2

| | |
|---|---|
| Phase I | |
| Lantrol | 5.0 |
| Stearic acid | 15.0 |
| "Tegacin" reg. | 2.0 |
| Myristyl alcohol | 0.5 |
| Mineral oil (70 vis.) | 8.0 |
| Isoropyl palmitate | 3.0 |
| Phase II | |
| $H_2O$ | 55.4 |
| Triethanolamine | 1.0 |
| "Tegosept" | 0.1 |
| Propylene glycol | 10.0 |

Heat Phase I and Phase II to 78°C. Add Phase I to Phase II, agitate and cool to 50°C. Color and perfume, agitate to 32°C and fill.

No. 3

| | |
|---|---|
| Stearic acid | 140 |
| Lauryl alcohol | 30 |
| Triethanolamine | 7 |
| Sodium lauryl sulfate | 5 |
| Glycerin | 50 |
| Water | 770 |
| Preservative | 10 |
| Perfume | |

Mix the first three ingredients and heat to 180°F (82°C). Add the sodium lauryl sulfate and the glycerin to the water and heat to same temperature and add to previous mixture with stirring. Continue stirring to 140°F, adding preservative and perfume.

No. 4

Mentholated Lemon

| | |
|---|---|
| Stearic acid | 1 lb. |
| White beeswax | 2 oz. |
| Glycerin | 4 pt. 6 oz. |
| Morpholine | 2 oz. |
| Warm water | 4½ pt. |
| Powdered borax | ¼ oz. |
| Menthol | 2 oz. |
| Terpeneless imitation lemon oil | 2 oz. |

Melt the first three ingredients in a large double boiler or on a water bath. When the waxes are melted, add the morpholine. Stir for about 10 min to get a good saponification; then add to this mixture a solution made of the borax and warm water. Stir again vigorously for 10 min, let cool, let stand for 10 min, then stir in a mixture of the methanol and lemon oil. Mix well and let stand 24 hr to shrink before putting into jars or tubes.

This cream leaves the skin smooth and refreshed. It is an ideal powder base cream.

---

**Night Cream**

No. 1

| | |
|---|---|
| Mineral oil | 280 |
| Olive oil | 45 |
| Lanolin | 125 |
| Stearic acid | 40 |
| Spermaceti | 65 |
| Cetyl alcohol | 125 |
| Triethanolamine | 109 |
| Water | 400 |
| Preservative | 10 |
| Perfume | |

Heat water to 158°F adding triethanolamine. Heat first six ingredients together to same temperature and stir water solution into them. Continue stirring until mixture cools to 122°F, then stir in preservative and perfume.

No. 2

| | |
|---|---|
| Lanolin absorption base | 24.0 |
| Robane | 3.0 |
| Isopropyl myristate | 5.0 |
| Lanolin oil | 2.0 |
| Carolate | 3.0 |
| Propylene glycol | 5.0 |
| Preservative | 0.2 |
| Water and perfume q.s. | 100.0 |

---

**Soft Skin Cream**

| | |
|---|---|
| Phase I | |
| Lantrol | 50.40 |
| Cetyl alcohol | 0.25 |
| Stearic acid | 2.00 |
| "Tween" 61 | 2.00 |
| "Arlacel" 20 | 1.00 |
| Hexachlorophene | 0.05 |
| Phase II | |
| "Carbopol" 934 (3% solution) | 10.00 |
| Triethanolamine (85%) | 0.28 |
| Phase III | |

| | |
|---|---|
| Water | 26.92 |
| Triethanolamine (85%) | 0.75 |
| Diethanolamine | 0.16 |
| Glycerine | 5.00 |
| Dehydroacetic acid | 0.16 |
| Sodium hydroxide | 0.03 |
| Sodium lauryl sulfate | 1.00 |

Heat all phases to 78°C. Add I to II slowly with agitation. Add I and II to III. Stir gently till 32°C. Mill Eppenbach, Setting 28.

---

**Heavy Cream (Ointment)**

| | |
|---|---|
| Cetyl alcohol | 3.5 |
| Stearyl alcohol | 7.0 |
| Sodium lauryl sulfate | 2.0 |
| "Robane" | 8.5 |
| Sesame oil | 5.0 |
| Glycerine | 5.0 |
| Preservative | 0.2 |
| Water and perfume q.s. | 100.0 |

---

**Emulsifier-free Moisturizing Cream**

| | |
|---|---|
| A | |
| "Veegum" | 5.0 |
| Water | 75.0 |
| B | |
| Propylene glycol | 2.0 |
| "Modulan" | 2.0 |
| Petrolatum, white | 10.0 |
| "American" L-101 | 6.0 |
| Preservative | q.s. |

Add the "Veegum" to the water slowly, continually agitating until smooth. Add B to I and heat to 70°C. Mix until uniform.

---

**Moisturizing Cream**

No. 1

| | |
|---|---|
| Phase I | |
| "Veegum" | 1.50 |
| Water | 74.20 |
| "Tegosept" M | 0.20 |
| Phase II | |
| Glycerin | 4.00 |
| Triethanolamine | 1.00 |
| Phase III | |
| "Tegosept" P | 0.10 |
| "Lantrol" | 10.00 |
| Stearic acid | 2.00 |
| Isopropyl myristate | 3.00 |
| Cetyl alcohol | 2.00 |

Add "Veegum" to $H_2O$, slowly agitating continually until smooth. Add II to I. Heat to 70°C. Heat III 78°C. Add slowly to I and II mixture with agitation. Cool to 50°C. Perfume and color. Agitate to 30°C and fill.

| | No. 2 Cream | No. 3 Lotion |
|---|---|---|
| Phase I | | |
| Beeswax | 7.50 | 7.50 |
| "Lantrol" | 5.00 | 5.00 |
| "Arlacel" 80 | 1.50 | 1.50 |
| "Arlacel" 85 | 0.50 | 0.50 |
| Petrolatum | 10.50 | — |
| Mineral oil, 70 vis. | 25.00 | — |
| Mineral oil, 185 vis. | — | 35.50 |
| P-"Parasept" | 0.05 | 0.05 |
| Phase II | | |
| Propylene glycol | 5.00 | 5.00 |
| Water | 44.14 | 44.14 |
| Borax | 0.66 | 0.66 |
| M-"Parasept" | 0.15 | 0.15 |

Heat Phase I and II to 78°C. Add Phase II to I slowly with agitation. Stir down to 38°C, and pass through Eppenbach colloid mill at Setting 28.

No. 4
(Penetrating)

| | |
|---|---|
| Phase I | |
| Water | 33.55 |
| Triethanolamine | 1.00 |
| "Veegum" Mix (5%) | 5.25 |
| Borax | 0.20 |
| "Tween" 20 | 0.20 |
| Phase II | |
| Polyethylene glycol 400 monostearate | 5.25 |
| "Tegin" 515 | 1.00 |
| Mineral oil 185 Vis. | 36.00 |
| "Lantrol" | 10.00 |
| Beewax | 4.20 |
| Myristyl alcohol | 0.50 |
| Orthophenyl phenol (purified) | 0.15 |
| "G 11" | 0.20 |
| Palmitic acid | 2.50 |
| "Veegum" mix | |

| | |
|---|---|
| (1) "Veegum" | 25.00 |
| Prop. glycol | 75.00 |
| (5) Orthophenyl phenol (purified) | 0.10 |
| Water | 400.00 |

Heat Phase I and II 78°C. Add I to II slowly with agitation. Cool to 50°C. Perfume and color agitate to 30°C. Fill.

### Hand Protective Cream

| | No. 1 | No. 2 | No. 3 |
|---|---|---|---|
| Triple-pressed stearic acid | 50.0 | 8.5 | 4.0 |
| Polyethylene glycol 1450 | 60.0 | — | 4.5 |
| Polyethylene glycol monostearate | — | 4.3 | — |
| Anhydrous lanolin | 20.0 | 1.0 | 1.5 |
| Terpineol | 0.3 | 0.1 | 0.1 |
| Triethanolamine | 4.0 | — | — |
| Potassium hydroxide | 2.0 | 0.5 | 0.2 |
| Propylene glycol | 36.0 | 5.0 | 7.4 |
| Hydroxyethyl cellulose | 45.0 | 15.0 | 7.0 |
| Perfume | 0.3 | 0.4 | 0.4 |
| Zinc stearate | 21.0 | — | 1.6 |
| Water | 200.0 | 65.2 | 73.0 |

A paddle-type stirrer is preferred for No. 1 to prevent aeration, but a higher-speed, mechanical mixer can be used to make the liquid creams (No. 2 and 3).

Melt the stearic acid, PEG-1450, monostearate, and lanolin and bring the temperature to 70°C. Add the terpineol. Dissolve the potassium hydroxide in about an equal weight of water, add this solution and the triethanolamine (where used) to the melt, and stir until most of the soap is in solution. Bring the temperature to 70°C. Heat the water to 70°C and stir into the mixture or stir the mixture into the water, if more convenient. Continue stirring until a homogeneous mixture is obtained. Add the propylene glycol and hydroxyethyl cellulose and continue stirring slowly until the cream has cooled to about 40°C. Stir in the zinc stearate and the perfume.

The cream can be packaged at 40°C as it has a better consistency than when allowed to cool to room temperature before packaging. The liquid creams (formulas No. 2 and 3) should be stirred slowly, but continuously, until cooled to room temperature to prevent increased viscosity on standing. The stirring of any of the creams should be fast enough to maintain thorough mixing without aerating the cream.

The creams develop a pearly texture on standing. The lanolin content may be decreased or omitted to increase the pearly texture.

No. 3
(With Lanolin)

| | |
|---|---|
| Polyethylene glycol 400 monostearate | 40 |
| Polyethylene glycol stearate | 50 |
| Lanolin (anhydrous) | 10 |
| Terpineol | 2 |
| Polyethylene glycol 1000 | 20 |
| Preservative | 10 |
| Water | 910 |
| Perfume and color | |

Heat first three ingredients together until melted, add the terpineol, and continue heating to 140°F. Heat water to same temperature and add it rapidly, with vigorous stirring. Then add the polyethylene glycol 1000 with slow stirring. When mixture has cooled to 104°F, add preservative and perfume. Continue stirring until product has cooled to room temperature.

No. 4
(Hand Cream)

| | |
|---|---|
| Phase I | |
| Water | 72.7 |
| "Sorbo" | 5.0 |
| "Tegosept" M | 0.2 |
| "Duponol" WA paste | 1.0 |
| Phase II | |
| "Lantrol" | 5.0 |
| Myristyl alcohol | 1.0 |
| "Tegin" 515 | 10.0 |
| "Tegosept" P | 0.1 |
| Isopropyl myristate | 5.0 |

Heat oil and water phases separately to 78°C. Add Phase I to Phase II with steady agitation. Stir down to about 50°C and perfume. Agitate to 30°C and fill.

---

**Four-Purpose Cream**

This cream has a cooling effect on the skin and is nongreasy. It serves equally well as a cleansing and as a massage cream. No oil or grease is left behind after the usual hot-towel application.

| | | |
|---|---|---|
| A. | White petrolatum | 4.5 lb. |
| | Paraffin wax | 18.0 |
| | White mineral oil | 18.0 |
| | Glycostearin | 18.0 |
| B. | Water | 71.0 lb. |
| | Triethanolamine | 2.0 |
| | Mold inhibitor* | 3.0 oz. |
| C. | Perfume | as req. |

Heat A to 166°F and stir until complete solution is obtained. Heat B to 166°F and add A to B, stirring continuously while cooling. Add C at 135°F and pour at 120 to 125°F.

*A preservative, parahydroxybenzoic acid derivative, generally used in the proportion of 18 oz to 100 gal. of the finished product. It should be dissolved by means of heat in the water called for in the formula.

## Hand Cream

No. 1

| | | |
|---|---|---|
| A. | *Oil Phase* | |
| | "Neodol" 25 | 4.5 |
| | Glyceryl monostearate | 2.5 |
| | Stearic acid | 1.5 |
| | Propyl p-hydroxybenzoate | 0.1 |
| B. | *Water Phase* | |
| | Triethanolamine (99%) | 0.3 |
| | Glycerine, USP | 5.0 |
| | Methyl p-hydroxybenzoate | 0.2 |
| | Distilled water | 85.9 |

Heat both phases separately to 70°C. Add water phase to oil phase with careful agitation. Allow mixture to cool with constant stirring.

No. 2

| | |
|---|---|
| Glyceryl monostearate (S.E.) | 4.0 |
| Stearic acid | 4.0 |
| Cetyl alcohol | 2.0 |
| Lanolin | 2.0 |
| "Robane" | 4.0 |
| Propylene glycol | 3.0 |

| | |
|---|---|
| Triethanolamine | 1.0 |
| Preservative | 0.2 |
| Water and perfume q.s. | 100.0 |

No. 3

| | |
|---|---|
| Yellow ceresin | 2.0 oz. |
| Yellow beeswax No. 1 | 2.0 |
| Stearic acid | 2.0 |
| White petrolatum | 4.0 |
| White mineral oil | 8.0 fl. oz. |
| Water | 6.0 |
| Borax | 120.0 gr. |
| Triethanolamine | 0.5 fl oz. |
| Perfume | to suit |

Melt the ceresin, beeswax, petrolatum, stearic acid and white mineral oil together, heating to 160°F. Dissolve the borax in the water by means of heat, and add the triethanolamine to the solution, bringing the temperature to 160°F. Pour the water solution all at once into the melted wax mixture and stir thoroughly. Remove from heat and continue stirring thoroughly until it begins to thicken; then add the perfume and mix well. Pour into jars, leaving the lids off until the cream is completely cold. From 6 to 8 oz. of good artificial rose oil to each 100 lb of cream will give a pleasing odor.

This fine cream is intended for regular use as a powder base as well as a massage and night cream. It is entirely harmless to the skin, relieves roughness, redness, and chapping, and will keep the skin smooth and soft.

---

**Foundation Cream**

| | |
|---|---|
| Glycerol monostearate | 35 |
| Stearic acid | 240 |
| Lanolin | 15 |
| Isopropyl myristate | 40 |
| Triethanolamine | 10 |
| Water | 800 |
| Aluminum hydroxide | 25 |
| Propylene glycol | 25 |

Heat the water and triethanolamine to 176°F. Heat the first four ingredients to the same temperature, and add the water solution to them with stirring. Continue stirring until temperature has fallen to

149°F; then add a mixture of the glycol and aluminum hydroxide.

---

**Emollient Cream**

| | |
|---|---|
| Phase I | |
| "Lantrol" | 10.00 |
| Stearic acid | 1.00 |
| Mineral oil, 70 visc. | 20.00 |
| Myristyl alcohol | 0.20 |
| Beeswax | 1.00 |
| "Span" 60 | 2.50 |
| Glyceryl monostearate P | 2.50 |
| "Tegosept" P | 0.05 |
| Phase II | |
| "Tween" 20 | 0.20 |
| Borax | 0.05 |
| Glycerin | 5.00 |
| Triethanolamine | 0.40 |
| Water | 56.50 |
| "Tween" 60 | 0.50 |
| "Tegosept" M | 0.10 |

Heat Phase I and II to 78°C. Add Phase I to II with agitation, cool to 50°C, perfume, agitate, and cool to 30°C. Fill.

---

**Face Cream**

No. 1

| | |
|---|---|
| Stearyl alcohol | 3.5 |
| Cetyl alcohol | 3.5 |
| Sodium lauryl sulfate | 1.0 |
| Robane | 5.0 |
| Stearic acid | 10.0 |
| Glycerin | 6.0 |
| Triethanolamine | 0.5 |
| Borax | 0.1 |
| Preservative | 0.2 |
| Water and perfume q.s. | 100.0 |

No. 2

With Lanolin

| | |
|---|---|
| Mineral oil | 570 |
| Beeswax | 25 |
| Lanolin | 60 |
| Stearic acid | 65 |
| Triethanolamine | 10 |
| Glycerin | 10 |
| Borax | 10 |
| Water | 400 |
| Preservative | 10 |
| Perfume | |

Mix the first four ingredients and heat to 167°F. Dissolve in water the glycerin, borax, and triethanolamine, with heating to same temperature; then add to first mixture with stirring. After cooling to 122°F, stir in perfume and preservative.

No. 3

| Biocidal | |
|---|---|
| Mineral oil | 47.9 |
| Beeswax | 6.0 |
| Spermaceti | 6.0 |
| Cetyl alcohol | 1.0 |
| Lanolin | 1.0 |
| Water | 38.0 |
| Cetyltrimethyl ammonium cyclehexyl sulfamate | 0.1 |
| Perfume | to suit |

**Day Cream**

| Cetyl alcohol | 3.5 |
|---|---|
| Stearyl alcohol | 7.0 |
| Sodium lauryl sulfate | 2.0 |
| Robane | 8.5 |
| Sweet almond oil | 5.0 |
| Glycerin | 5.0 |
| Preservative | 0.2 |
| Water and perfume q.s. | 100.0 |

**Cold Cream**

No. 1

| Stearic acid | 30 lb. |
|---|---|
| Anhydrous lanolin | 20 |
| White beeswax | 16 |
| Terpineol | 3 1/5 oz. |
| White mineral oil | 33 lb. |
| Triethanolamine | 4 |
| Propylene glycol | 16 |
| Water | 95 |

Melt the stearic acid lanolin, and beeswax in the mineral oil, heat to 70°C and then add the terpineol. Heat the water to 70°C in a separate kettle, add the triethanolamine, and then add this solution to the hot mixture of wax and oil. Stir vigorously until a creamy emulsion is obtained. Add the perfume to the propylene glycol and add this solution to the emulsion. Continue stirring until the emulsion is smooth and quite viscous, and then stir occasionally until room temperature is reached.

It is possible to pour this cream into jars while still warm and thin enough to pour, but the resulting cream may not have the smooth texture of a cream that is packaged when cold. A pressure filler is usually necessary to fill the containers with the emulsion at room temperature.

This is a smooth, stable cream that can be easily applied to the skin, even though it is more viscous than the cleansing creams. It is readily removed with a soft cloth or absorbent tissue, or it can be washed from the skin with water.

This formula should serve as a starting point for making a cold cream to suit the individual preference, and should not be considered as necessarily the best product obtainable. Great variation in the wax and oil constituents is possible with little change in the basic ingredients. For example, vegetable oils, such as sweet-almond or olive oil, may be substituted for all or part of the mineral oil to produce an excellent product.

No. 2

| | |
|---|---|
| "Carlate" | 12.5 |
| Beeswax, white | 12.0 |
| Sesame oil | 40.0 |
| Robane | 16.0 |
| Borax | 0.5 |
| Preservative | 0.1 |
| Water and perfume q.s. | 100.0 |

---

**All Purpose Cream**

No. 1

| | |
|---|---|
| Lauryl alcohol | 110 g. |
| Beeswax | 80 g. |
| Paraffin | 70 g. |
| Mineral oil | 260 g. |
| Sodium lauryl sulfate | 10 g. |
| Triethanolamine | 4 g. |
| Water | 465 ml. |
| Preservative | 10 ml. |
| Perfume | |

Heat first four ingredients to 180°F. Add sodium lauryl sulfate, preservative, and triethanolamine to water and heat to same temperature, adding to first mixture slowly with stirring. Continue stirring until mixture has cooled to 140°F, adding perfume.

| | No. 2 | No. 3 |
|---|---|---|
| A. *Oil Phase* | | |
| "Neodol" 45 | 23.0 | 24.4 |
| Paraffin (130°MP) | 4.0 | 4.0 |
| Mineral oil | 23.0 | 23.0 |
| White beeswax | 14.0 | 12.6 |
| Propyl p-hydroxybenzoate | 0.1 | 0.1 |
| B. *Water Phase* | | |
| Borax | 1.0 | 1.0 |
| Methyl p-hydroxybenzoate | 0.2 | 0.2 |
| Distilled water | 34.7 | 34.7 |

Heat phase A and B separately to 70°C. Add phase A to phase B with careful agitation. Allow mixture to cool to room temperature with constant stirring.

---

**Lotion Vehicle**

| | |
|---|---|
| A. *Oil Phase* | |
| "Amerchol" L-101 | 6.0 |
| "Modulam" | 2.0 |
| Stearic acid, XXX | 2.0 |
| Sorbitan monostearate | 2.1 |
| Polyoxyethylene (20) sorbitan monostearate | 2.9 |
| "Methocel" 65 HG, 4000 cps. 1 | 0.1 |

B. *Water Phase*

| | |
|---|---|
| Propylene glycol | 5.0 |
| Water | 79.9 |
| Perfume and preservative | q.s. |

Add the water phase at 80°C to the oil phase at 80°C with stirring. Mix while cooling to 40°C. Add perfume. Continue mixing and cool to 30°C.

---

**Lotions**

| | No. 1 | No. 2 | No. 3 |
|---|---|---|---|
| "Standapol" WAQ special | 50 | 30 | 10 |
| "Standapol" ES-2 | — | 20 | 40 |
| "Standamid" LD | 5 | 4 | 3 |
| "Monamid" S | — | 1 | 1 |
| "Aromox" DM 16 D | — | 2 | 1 |
| Ethylene glycol monostearate | 1 | 2 | 1 |
| Sodium chloride | 1 | — | 1 |
| Water, perfume, preservative, color, egg, milk, etc. | q.s. | q.s. | q.s. |

No. 4

| | |
|---|---|
| Glyceryl monostearate (S.E.) | 2.7 |
| Cetyl alcohol | 1.5 |
| Silicone 200 | 1.5 |
| Lanolin oil | 2.0 |
| "Robane" | 3.0 |
| Sodium lauryl sulfate | 0.3 |
| Preservative | 0.2 |
| Water q.s. | 100.0 |

No. 5

A. *Oil Phase*

| | |
|---|---|
| "Neodol" 45 | 4.5 |
| "Neodol" 23-3A | 2.5 |
| Stearic acid | 1.5 |
| Propyl p-hydroxybenzoate | 0.1 |

B. *Water Phase*

| | |
|---|---|
| Triethanolamine (99%) | 0.3 |
| Glycerin, USP | 5.0 |
| Methyl p-hydroxybenzoate | 0.2 |

| | |
|---|---|
| Distilled water | 85.9 |

Heat both oil and water phases separately to 70°C. Add oil phase to the water phase with careful agitation. When emulsion is well formed allow mixture to cool to room temperature with constant stirring.

No. 6

| | |
|---|---|
| Glycerol monostearate | 50 |
| Oleic acid | 30 |
| Mineral oil | 15 |
| Lanolin | 10 |
| Triethanolamine | 12 |
| Sodium lauryl sulfate | 10 |
| Preservative | 10 |
| Water | 980 |
| Color and perfume | |

Dissolve sodium lauryl sulfate and preservative in water and heat to 194°F. Heat mixture of first four ingredients to same temperature and add them slowly to 400 ml water, with stirring. Add triethanolamine to remainder of water, heat to 194°F, and stir into other mixture. Then stir in color and perfume.

No. 7

| | |
|---|---|
| "Modulan" | 2.0 |
| Lanolin | 10.0 |
| Stearic acid | 2.5 |
| Glyceryl monostearate | 3.5 |
| Glycerin | 5.0 |
| Triethanolamine | 1.0 |
| Ethyl alcohol | 4.0 |
| Water | 82.0 |
| Perfume and preservative | to suit |

Heat the fats and the water-soluble ingredients in separate vessels to 90°C. Add the aqueous solution to the fats while agitating slowly, and continue mixing until the batch cools. The alcohol should be added at about 45°C along with the perfume.

No. 8

| | | |
|---|---|---|
| A. | Lanolin | 2.0 dr. |
| | "Aquaphor" | 3.0 dr. |
| | Propylene stearate | 6.0 dr. |

| | | | |
|---|---|---|---|
| B. | Boiling water | | 10.0 oz. |
| | Triethanolamine | | 60.0 min. |
| C. | Sodium benzoate | | 8.0 gr. |
| | Titanium dioxide | | 2.0 dr. |
| | Tincture of benzoin | | 60.0 min. |
| | Glycerin | | 0.5 oz. |
| | Bentonite | | 1.0 dr. |
| | Water | to make | 1.0 pt. |
| | Perfume oil | | Sufficient |

Melt A in a double boiler. Add B and stir to a good emulsion. Then add C and run the whole mixture through a homogenizer. Let cool and then stir in the perfume oil. Let stand 24 to 48 hr to shrink before bottling.

This lotion is excellent for rough, dry skin, to soothe windburn and sunburn, after shaving and bathing, and may be used as a powder base. It rubs in without leaving a sticky after-use feeling.

No. 9
(Body)

| | |
|---|---|
| Phase I | |
| "Lantrol" | 7.00 |
| Isopropyl palmitate | 10.00 |
| Oleic acid | 2.70 |
| "Tegin" 515 | 4.00 |
| "Tegosept" P | 0.10 |
| DL TDP (antioxidant) | 0.01 |
| Phase II | |
| Triethanolamine | 1.30 |
| Water | 74.69 |
| "Tegosept" M | 0.20 |

Heat Phase I and II to 78°C. Add Phase I to Phase II with stirring (do not hemogenize). Cool to 50°C. Perfume and color. Cool to 30°C. Fill.

**Facial Astringent**

| | |
|---|---|
| Rose water | 6 fl.oz. |
| Witch hazel | 10 fl.oz. |
| Tannic acid | 10 gm. |

**Medicated Hospital Lotion**

| | |
|---|---|
| Menthol | 0.20 |
| Hexachlorophene | 0.10 |
| "Ritalan" | 1.00 |
| "Super-Sat" | 2.00 |
| Mineral oil lite, N. F. | 3.00 |

| | |
|---|---|
| Cetyl alcohol | 2.00 |
| Glycerin | 5.00 |
| "Super-Sat" AWS 3 | 1.00 |
| Methyl paraben | 0.10 |
| Propyl paraben | 0.05 |
| Water q.s. | 100.00 |
| Perfume q.s. to suit | |

Melt and blend Hexachlorophene, "Ritalan", "Super-Sat", mineral oil, cetyl alcohol, glycerin, "Super-Sat" AWS 3 and the two parabens. Add water at 85°. Stir to form emulsion and continue agitation to 40°. Dissolve menthol in 2 to 3 ml of 95% ethyl alcohol and add to emulsion. Perfume and agitate to room temperature.

---

**Fragrance Concentrates**

Concentrates are never used undiluted. To make a "skin" or handkerchief perfume, use 1 oz. of concentrate to 4 or 5 oz. of ethyl alcohol. For colognes use 1 oz. of concentrate to 12 to 15 oz. of alcohol; for creams and lotions, $\frac{1}{4}$ to $\frac{1}{2}$ oz. per gallon.

| Oriental Type (Chypre) | No. 1 | No. 2 | No. 3 |
|---|---|---|---|
| Oakmoss | 6.0 | 8.0 | 8.0 |
| Oil patchouly | 10.0 | 12.0 | 12.0 |
| Oil vetivert | 5.0 | 10.0 | 10.0 |
| Oil sandalwood E.I. | 10.0 | 10.0 | 10.0 |
| Oil bergamot | 15.0 | 15.0 | 10.0 |
| Isobutyl paracresol | 0.5 | 0.5 | 0.5 |
| Oil coriander | 2.0 | — | — |
| Isoamul acetate | 5.0 | 5.0 | — |
| Oil sassafrass | 1.0 | — | — |
| Oil cassia | — | 5.0 | 5.0 |
| Oil angelica root | — | 3.0 | 3.0 |
| Geraniol | — | 5.0 | 5.0 |
| Heliotropine | 2.0 | — | — |
| Isoeugenol | 1.5 | — | 2.0 |
| Phenyl ethyl alcohol | 20.0 | 10.0 | 5.0 |
| Musk ketone | 15.0 | 15.0 | — |
| Oleoresin orris root | — | — | 10.0 |
| Musk xylol | 7.0 | — | 15.0 |
| Balsam Peru | — | — | 5.0 |

| Lilac concentrate | |
|---|---|
| Terpineol | 12.0 |
| Heliotropine | 25.0 |
| Cinamic alcohol | 5.0 |

| Violet concentrate | |
|---|---|
| Phenyl ethyl alcohol | 20.0 |
| Geraniol | 3.0 |
| Alpha ionone | 50.0 |

| | |
|---|---|
| Isobutyl paracresol | 0.5 |
| Phenyl ethyl alcohol | 19.5 |
| Linalool | 10.0 |
| Alpha ionone | 8.0 |
| Hydroxycitrochellol | 5.0 |
| Isoamyl acetate | 5.0 |
| Musk ketone | 10.0 |
| Phenyl acetaldehyde | 2.0 |
| Hydroxycitronellol | 7.0 |
| Isoamyl acetate | 5.0 |
| Musk xylol | 14.0 |

Carnation concentrate

| | |
|---|---|
| Geraniol | 2.0 |
| Hydroxy citronellol | 5.0 |
| Amyl cinamic aldehyde | 4.0 |
| Terpineol | 8.0 |
| Eugenol | 10.0 |
| Isoeugenol | 17.0 |
| Linalool | 5.0 |
| Methyl ionone | 5.0 |
| Indol (10% solution) | 2.0 |
| Benzyl acetate | 4.0 |
| Musk ambrette | 5.0 |
| Musk ketone | 5.0 |
| Methyl anthrinalate | 3.0 |

Rose concentrate

| | |
|---|---|
| Geraniol | 30.0 |
| Phenyl ethyl alcohol | 25.0 |
| Benzyl acetate | 5.0 |
| Isoamyl acetate | 5.0 |
| Isobutyl paracresol | 0.5 |
| Ionone | 10.0 |
| Amyl cinamic aldehyde | 2.5 |
| Phenyl acetaldehyde | 2.0 |
| Musk xylol | 20.0 |

**Perfume for Lotion**

| | |
|---|---|
| Bergamont | 29 |
| Benzaldehyde | 41 |
| Lemon oil | 10 |
| Clove oil | 10 |
| Mace oil | 10 |

**Itch Reliever**

| | |
|---|---|
| Talc | 90.0 |
| Undecylenic dialkylolamide | 5.0 |
| Menthol | 0.5 |
| Benzocaine | 0.1 |
| Perfume | to suit |

Add water and triturate to make a smooth paste.

**Hair Cream**

No. 1

(Cream)

Phase I

| | |
|---|---|
| "Lantrol" | 5.00 |
| Mineral oil 355 vis | 12.00 |
| Castor oil | 10.00 |

| | |
|---|---|
| Petrolatum | 5.00 |
| "Tegosept" P | 0.10 |
| Phase II | |
| 2 1/2% solution "Carbopol" 934 | 40.00 |
| Propylene glycol | 9.00 |
| Glycerine | 2.50 |
| "Brij" 35 SP | 1.00 |
| "Tegosept" M | 0.20 |
| Phase III | |
| Water | 14.24 |
| KOH | 0.96 |

Heat I, II, and III to 78°C. Add I to II. Agitate. Add III, cool to 50°C. Perfume. Agitate, cool to 32°C. Mill Eppenbach, setting 28.

No. 2

| | |
|---|---|
| Phase I | |
| Mineral oil, 185 vis. | 30.39 |
| "Lantrol" | 5.00 |
| 170/175 (micro crystalline wax) | 2.00 |
| Beeswax | 3.00 |
| "Tegosept" P (Preservative) | 0.10 |
| Phase II | |
| Water | 48.66 |
| Borax | 0.15 |
| "Tegosept" M (Preservative) | 0.20 |
| Phase III | |
| Water | 10.00 |
| Calcium, saccharated | 0.50 |

Heat I and II to 78°C. Add II to I, agitate. Add III, heated to 50°C. Agitate to 50°C. Perfume, agitate to 32°C. Mill Eppenbach, setting 28.

No. 3

| | |
|---|---|
| Diglycol stearate | 7.0 |
| "Robane" | 20.0 |
| Preservative | 0.2 |
| Water and perfume, q.s. | 100.0 |

No. 4
(Conditioning)

| | |
|---|---|
| Phase I | |
| Water | 46.70 |
| Glycerine | 3.00 |
| "Duponol" WA paste | 1.00 |
| "Tegosept" M | 0.20 |
| Phase II | |
| Isopropyl myristate | 8.00 |
| "Nimlesterol" D | 10.00 |
| Mineral oil, 70 vis. | 8.00 |
| Cetyl alcohol | 2.00 |
| "Tegin" 515 | 8.00 |
| "Brij" 35 | 8.00 |
| Microcrystallinewax, 155°F. | 5.00 |
| "Tgosept" P | 0.10 |

Heat phase I and II to 78°C. Add II to I with agitation, perfume and color at 50°C. Agitate, cool to 30°C. Fill.

| (Pomade) | No. 5 | No. 6 |
|---|---|---|
| "Lantrol" | 5.00 | — |
| "Lanfrax" | — | 30.00 |
| Mineral oil, 355 vis. | 55.00 | 55.00 |
| Isopropyl myristate | 15.00 | 10.00 |
| "Multiwax" 445 | 20.00 | — |
| Polyethylene glycol 400 dioleate | 5.00 | 5.00 |

Heat all ingredients to 78°C. Stir down to 50°C and perfume. Stir down to 5° above solidifying point and fill.

**Pomades—Anhydrous, Washable**

| | No. 6 | No. 7 | No. 8 |
|---|---|---|---|
| Polyethylene glycol 400 monostearate | 5% | 5.0% | — |
| Polyethylene glycol 400 dilaurate | — | — | 5 |
| Petrolatum USP white | 50 | 82.5 | 65 |
| Mineral oil, 180 vis | 30 | — | — |
| Lanolin alcohols ethoxylate (5 mole EO) | — | 2.5 | — |
| Lanolin alcohols ethoxylate | | | |

| | | | |
|---|---|---|---|
| (25 mol EO) | — | 2.5 | — |
| Acetylated lanolin alcohols (liquid fraction) | — | — | 10 |
| Isopropyl lanolate | 5 | — | 5 |
| Acetylated lanolin | 10 | — | — |
| Hexadecyl alcohol | — | — | 15 |
| Lanolin fatty acids (4) | — | 7.5 | — |

Stir slowly while heating to 85°C. Continue stirring and hold at this temperature until mixture is uniform. Cool to 40 to 45°C while stirring. Add perfume and fill just above the set point.

No. 9
(Butch Stick)

| | |
|---|---|
| "Lanfrax" | 25.0 |
| Isopropyl palmitate | 10.0 |
| White petrolatum | 35.0 |
| Microcrystaline wax, 190—195°F. | 20.0 |
| "Tween" 80 | 10.0 |
| Perfume | q.s. |

Heat all ingredients to 78°C or until all waxes are melted and blend thoroughly with agitation. Cool with stady agitation to 5 or 10° above the solidification point and perfume. Pour into suitable molds.

No. 10
(Liquid)

| | |
|---|---|
| White mineral oil | 1 gal. |
| Purified or medicinal grade oleic acid | 16 fl. oz. |
| Triethanolamine | 8 fl. oz. |
| Karaya gum | 2 oz. |
| Oxyquinoline sulfate | ¼ oz. |
| Water | 4 gal. |
| Perfume | to suit |

Mix the oleic acid with a half gal. of the white mineral oil, stirring until completely uniform. To this mixture, add the triethanolamine and continue to stir until a clear viscous mixture is formed. Then stir the remaining half gal. of white mineral oil into the mixture. Soak the karaya gum overnight in 1 gal. of cold water in which the oxyquinoline sulfate has been dissolved. In the morning, stir until it forms a perfectly even mixture. Stir the gum solution into the oleic acid-triethanolamine mixture. A white emulsion will immediately form.

Agitate thoroughly to form a perfectly smooth emulsion.

Then add the remaining 3 gal. of water gradually, with constant agitation, to the heavy emulsion first formed. This amount of water may be added without any danfer of the emulsion separating on standing. Less water may be added if a heavier product is desired. The karaya gum is added as a stabilizer for the emulsion, and the oxyquinoline sulfate as an antiseptic and antidandruff ingredient.

Perfume to suit with any scenting oil, using just enough to give the scent desired. About 2 oz. of perfume oil to 5 gal. of finished product will be sufficient.

Either the medicinal grade of white mineral oil or a technical grade, such as is used in the manufacture of cold creams, may be used.

A homogenizer or an electrically driven agitator should be used for mixing.

No. 11

(Liquid)

The following is an essentially nongreasy, clear preparation that leaves the hair soft, yet holds it in place:

| | |
|---|---|
| PVC K-30 | 1.0 |
| "Emulphor" EL 719 | 1.5 |
| "Emulphor" ON-670 | 0.5 |
| Ethanol (SDA-40) | 15.0 |
| Water | 82.0 |

Dissolve PVP in ethanol-water mixture. Add the other ingredients and shake mixture to yield a clear solution. If desired, a bacteriostat, such as hexachlorophene, and a modifier, such as a water-souble lanolin derivative, may be added to enhance the performance of the product.

No. 12

(Quick Breaking Foam Type)

| | | |
|---|---|---|
| "Polawax" | | 1.5 |
| PVP/VA E-735 | | 2.5 |
| 2-Ethylhexanediol-1,3 | | 5.0 |
| Dipropylene glycol | | 2.5 |
| "Acetulan" | | 1.0 |
| Water | | 35.0 |
| Perfume | | q.s. |
| Above concentrate | 92% | |
| Propellant 12/114 (20/80) | 8% | |

## Hair Straighteners

No. 1

(Caustic Type)

| | |
|---|---|
| Gum tragacanth | 2.0 |
| Boric acid | 1.0 |
| Water | 40.0 |

Make a uniform paste and stir in previously dissolved:

| | |
|---|---|
| Sodium carbonate | 1.0 |
| Potassium hydroxide | 1.0 |
| Glycerin | 2.0 |
| Water | 8.0 |

Apply the paste to hair with combing. Allow to remain for ½ hour. Then wash well with water to remove all paste from the hair.

No. 2

(Thio Type)

| | |
|---|---|
| Water | 2400.0 |
| Sodium hydroxide | 24.0 |
| Ethylene glycol | 15.0) slurry |
| CMC 70, high v sc. | 7.5) |
| Ammonia 26°B | 74.0 |
| Ammonium thioglycolate 60% | 324.0 |
| Perfume | 26.0 |

No. 3

| | |
|---|---|
| Water | 1000.0 |
| "Veegum" | 40.0 |
| Lauryl monoethanolamide | 20.0 |
| Ammonium thioglycolate | 205.0 |
| Ammonia 20°B | 52.0 |
| Sodium hydroxide | 12.0 |

## Hair Wave-Set

No. 1

| | |
|---|---|
| Gelatin | 10 |
| Water | 800 |
| Sodium lauryl sulfate | 109 |
| Ethyl alcohol | 20 |
| Propylene glycol | 90 |
| Preservative | 20 |

Perfume and color.

Add 500 parts of the water, heated to 140°F to the gelatin and soak until uniform solution is produced. Dissolve the sodium lauryl sulfate in the remaining water and add to gelatin solution. Mix the other ingredients separately and add to above solution.

No. 2

| | |
|---|---|
| Karaya gum | 10 |
| Isopropyl alcohol | 200 |
| Preservative | 10 |
| Water | 800 |
| Propylene glycol | 25 |
| Potassium carbonate | 3 |

Perfume and color.

Dissolve the gum in the alcohol, add the preservative, add 650 ml of the water, heat to 140°F, let stand for several hours, then filter. Dissolve the potassium carbonate in the remaining water and add to solution. Then add remaining ingredients.

**Permament Wave**

No. 1

| | |
|---|---|
| Potassium hydroxide | 10 |
| Sodium sulfite | 20 |
| Monoethanolamine | 50 |
| Sulfonated castor oil | 15 |
| Water | 900 |

Add the potassium hydroxide to 100 ml of the water, and stir until *completely* dissolved. Then dissolve the sodium sulfite in another 100 ml of the water. Mix the sulfonated castor oil and monoethanolamine. Then mix the three solutions and add the remaining water.

This is a strong solution. Be sure ingredients are dissolved, and instruct user to wash off carefully after hair is waved.

No. 2

| | | |
|---|---|---|
| Triethanolamine | | 16.0 fl. oz. |
| Sodium hyposulfite | | 6.0 oz. |
| Water | to make | 1.0 gal. |
| Perfume | | to suit |

Dissolve the sodium hyposulfite in a half gal. of water. Add the triethanolamine and then make up to 1 gal. with water. Add any desired perfuming oils.

This is an ideal product where a soft, deep wave instead of a tight curl is wanted.

No. 3

| | | |
|---|---|---|
| Powdered sodium carbonate | | 8.0 oz. |
| Powdered sodium hyposulfite | | 6.0 oz. |
| Distilled water | to make | 1.5 gal. |
| Morpholine | | 6.0 fl. oz. |
| Sulfonated castor oil | | 1.0 fl. oz. |
| Perfume oil and color | | to suit |

Dissolved the sodium carbonate and sodium hyposulfite separately, each in 3 pints of water. Add the morpholine to the sodium carbonate solution. Then mix the two solutions and stir in the sulfonated castor oil, which has been previously mixed with the perfume oil. Color to suit with a trace of caramel or certified cosmetic color. Do not over-color.

This waving fluid is suitable for both spiral and croquignole waves. It contains no caustic alkalies and may be safely used on all colors and qualities of hair.

---

**Hair Bleach**

| | |
|---|---|
| Stearyl sulfate | 20.0 |
| Water | 40.0 |
| Hydrogen peroxide | 30.0 |

---

**Aerosol Hair Spray**

No. 1

| | |
|---|---|
| Hair spray concentrate | 50.0 |
| "Aerothene" MM solvent | 12.5 |
| Propellant 11 | 12.5 |
| Propellant 12 | 25.0 |
| Perfume | to suit |

No. 2

| | |
|---|---|
| Hair spray concentrate | 50.0 |
| "Aerothene" MM solvent | 20.0 |
| "Aerothene" TT solvent | 12.0 |
| Propellant 12 | 16.0 |
| Nitrous oxide (to 60 psig) | 2.0 |
| Perfume | to suit |

No. 3

| | |
|---|---|
| Diisopropyl adipate | 20.0 |
| "Purcellin" | 10.0 |
| Isopropyl myristate | 10.0 |
| Paraffin oil 5°C | 28.0 |
| Chlorothene | 20.0 |
| Isopropyl alcohol | 10.0 |
| Perfume | 1.0 |
| "Katioran" SK | 1.0 |

The aerosol is loaded with 40% concentrate and 60% propellant

11/12 (50:50).

No. 4
(Gentle)

| | |
|---|---|
| PVP K-30 (100% solids) | 1.00 |
| "Ethoxylan" 100 | 0.40 |
| Silicone (DC-556) | 0.25 |
| Ethanol (SDA-40, anhydrous) | 28.35 |
| Propellant 11/12 (65/35) | 70.00 |
| Perfume | q.s. |

No. 5
(Regular)

| | |
|---|---|
| PVP/VA E-635 (50% solids) | 4.80 |
| Dimethyl phthalate | 0.25 |
| Silicone (DC-556) | 0.10 |
| Ethanol (SDA-40, anhydrous) | 24.70 |
| Propellant 11/12 (65/35) | 70.00 |
| Perfume | q.s. |

No. 6
(Hard-to-Hold)

| | |
|---|---|
| "Gantrez" ES-225 | 3.96 |
| Ethanol (SDA-40, anhydrous) | 25.02 |
| Triisopropanolamine | 0.07 |
| Propellant 11 | 45.00 |
| Propellant 12 | 25.00 |
| Perfume | q.s. |

No. 7
(Protein-Type)

| | |
|---|---|
| PVP/VA E-635 or E-735 | 2.00 |
| Polypeptide AAS, 25% | 1.00 |
| "Uvinul" D-50 | 0.50 |
| Silicone (DC-556) | 0.10 |
| Methylene chloride | 10.00 |
| Ethanol (SDA-40, anhydrous) | 41.65 |
| Propellant 11/12 (60/40) | 45.00 |
| Perfume | q.s. |

Additions of "FC-3569" ranging from 0.0082 to 0.05% improve

the moisture resistance and film forming properties of hair sprays and add to the combability of the sprayed hair.

### Wig Luster Spray

| | No. 1 | No. 2 |
|---|---|---|
| PVP/VA E-635 (50% solids) | 4.80 | — |
| PVP/VA E-535 (50% solids) | — | 3.5 |
| Dimethyl phthalate | 0.25 | 0.1 |
| Silicone | 0.10 | 0.1 |
| "Ethylan," 35% | — | 0.5 |
| Ethanol (SDA-40, anhydrous) | 24.70 | 25.6 |
| Propellant 11/12 (65/35) | 70.00 | 70.0 |
| Perfume | 0.15 | 0.2 |

### Depilatories

No. 1

| | |
|---|---|
| Strontium sulfide | 35.0 |
| Titanium dioxide | 21.0 |
| Corn starch | 43.0 |
| Benzocaine | 0.2 |
| Perfume | 0.8 |

No. 2

| | |
|---|---|
| Strontium sulfide | 40.0 |
| Zinc oxide | 15.0 |
| Powdered soap | 5.0 |
| Corn starch | 20.0 |
| Chalk | 20.0 |

For application the above are mixed with water into a paste of suitable consistency.

No. 3

Paste Depilatory

| | |
|---|---|
| Sodium sulfide | 4.0 |
| Calcium hydroxide | 4.0 |
| Glycerin | 1.0 |
| Kaolin | 32.0 |
| Water | 59.0 |

No. 4

Cream Depilatory

| | |
|---|---|
| Strontium sulfide | 30.0 |
| Titanium dioxide | 10.0 |
| Sorbitol syrup | |
| (Neutral) sodium | 10.0 |
| 42.5° Be | 0.5 |
| Methyl cellulose | 2.5 |
| Demineralized water | 47.0 |

No. 5

| | |
|---|---|
| Calcium thioglycolate | 6 |
| Strontium hydroxide powder (monodydrate) | 5 |
| Polyethylene glycol 6000 | 9 |
| "Kelzan" | 2 |
| Glycerin | 10 |
| Zinc oxide | 10 |
| Chalk | 5 |
| Water | 52 |
| Perfume | 1 |

No. 6

| | |
|---|---|
| Calcium thiglycolate | 150 |
| Calcium hydroxide | 50 |

| | |
|---|---|
| Calcium carbonate | 600 |
| Water | 500 |
| Sodium lauryl sulfate | 5 |
| "Lanette" wax | 120 |

---

**Pre-Shave Lotion, Electric**

No. 1

| | |
|---|---|
| Phase I | |
| "Lantrol" AWS | 2.50 |
| Hexachlorophene | 0.10 |
| Oleyl alcohol | 2.50 |
| Isopropyl myristate | 5.00 |
| Dipropylene glycol | 4.00 |
| Alcohol SDA 40 anhydrous | 70.80 |
| Phase II | |
| Allantoin | 0.10 |
| Water | 15.00 |

Add ingredients of Phase I to mixing vessel along with perfume. Agitate till solution forms. Do same with Phase II, and very slowly add Phase II to I with agitation. Color and fill.

No. 2

| | |
|---|---|
| Ethyl alcohol, SD anhydrous | 94.76 |
| Hexachlorophene | 0.3 |
| "RC-3569" | 0.64 |
| Propylene glycol | 3.00 |
| Perfume | 1.30 |

No. 3

| | |
|---|---|
| Hexachlorophene | 1.5 |
| Alcohol | 400.0 |
| Propylene glycol | 50.0 |
| Water | 550 |
| Perfume | |

Dissolve the hexachlorophene in 20 ml. of the alcohol, then add the remainder of the alcohol. Add propylene glycol to the water and stir into the alcoholic solution.

---

**Lather Shave Cream**

| | No. 1 | No. 2 | No. 3 | No. 4 |
|---|---|---|---|---|
| "Neo-Fat" 265 | 7.86 | | | |
| "Neo-Fat" 255 | | 13.50 | 14.00 | 14.00 |
| "Neo-Fat" 16-54 | | | 7.00 | 7.00 |
| "Neo-Fat" 18-54 | | | 7.00 | 7.00 |
| "Neo-Fat" 18-58 | | 31.50 | | |
| "Neo-Fat" 18-55 | 36.03 | | | |
| 50% Potassium hydroxide | 16.00 | 15.32 | 10.04 | 9 32 |
| 50% Sodium hydroxide | 2.00 | 1.92 | | |
| "Aromox" c/12-W | 1.00 | 1.00 | 1.00 | 1.00 |
| Glycerin | 2.17 | 5.00 | 12.00 | 12.00 |
| Water | 34.94 | 31.76 | 48.96 | 49.68 |
| Perfume and preservatives | q.s. | q.s. | q.s. | q.s. |

Heat fatty acids, glycerin and 3/4 of the water to 75 to 80°C. Add alkalis and balance of water while mixing slowly. If soap gels form, mix throughly and set for 2 to 3 hr. Add "Aromox" C/12-W, at 40 to 45°C add perfume. On cooling adjust evaporation losses with water.

No. 5
(Mentholated)

| | |
|---|---|
| Coconut oil soap | 450.0 |
| Stearic acid | 50.0 |
| Sodium lauryl sulfate | 25.0 |
| Menthol | 2.5 |
| Alcohol | 10.0 |
| Triethanolamine | 10.0 |
| Glycerin | 170.0 |
| Water | 600.0 |

Add sodium lauryl sulfate to water and heat to 158°F. Dissolve the menthol in the alcohol, add the triethanolamine. Add the glycerin slowly, with stirring. Melt stearic acid separately and stir into mixture, continuing stirring until thoroughly mixed. Then add the perfume.

---

**Emulsifier-free Shaving Gel**

| | |
|---|---|
| A | |
| "Veegum" | 5.0 |
| Water | 82.0 |
| B | |
| "Americhol" L-101 | 2.5 |
| "Acetulan" | 0.5 |
| Silicone, 350 cs | 10.0 |
| Preservative | q.s. |

1. Add the "Veegum" to the water slowly, agitating continually until smooth.
2. Continue stirring, add B and mix until uniform.

Directions for use: Wash face thoroughly with soap and water. Leave face wet. Apply gel sparingly with fingers and shave.

**Brushless Shaving Cream**

No. 1

| | |
|---|---|
| Stearic acid | 180 |
| Mineral oil | 50 |
| "Surfonic" N-40 | 50 |
| Sorbitol | 50 |
| Borax | 20 |
| Triethanolamine | 10 |
| Preservative | 10 |
| Water | 640 |
| Perfume | |

Heat first three ingredients to 194°F. Mix water, sorbitol, borax, triethanolamine, and preservative solution; heat to 203°F. Add water solution to first mixture, with stirring. Continue stirring until it has cooled to set point and stir occasionally until product has cooled to room temperature, then add perfume.

Have exhaust fan or good ventilation for ammonia fumes.

No. 2

| | |
|---|---|
| Stearic acid | 150 |
| Propylene glycol monostearate | 35 |
| Mineral oil | 30 |
| Glycerin | 50 |
| Triethanolamine | 15 |
| Water | 900 |

Heat the first three ingredients to 176°F. Dissolve the triethanolamine and glycerin in the water, heat to the same temperature, and add this solution to the first mixture, slowly and with stirring.

No. 3
(Lanolinized)

| | |
|---|---|
| Triple-pressed stearic acid | 16.0 |
| White mineral oil | 4.0 |
| Anhydrous lanolin | 3.5 |
| Terpineol | 0.1 |
| Polyethylene glycol monostearate | 3.2 |
| Potassium hydroxide | 0.8 |
| Propylene glycol | 4.0 |
| Water | 68.1 |

Melt the stearic acid, mineral oil, lanolin, and monostearate; heat to 60°C and add the terpineol. In a separate container, add the potassium hydroxide to the water and heat to 60°C. Add the water solution to the oil solution, or vice versa, and stir until a smooth emulsion is formed.

Add the propylene glycol and continue stirring slowly, but continuously, until smooth, viscous cream is obtained. Then stir at intervals until the cream has cooled to 30 to 35°C. The cream should be covered when not being stirred, and rapid stirring should be avoided after the emulsion has begun to thicken to prevent undesirable aeration of the cream. When the cream has cooled to about 35°C, add the perfume.

The cream can be packaged when it has cooled to about 32°C. A softener cream is obtained when it is allowed to stand overnight before packaging.

---

**Nonseparating Hot Shave Formulation**

Sequence A

| | |
|---|---|
| "Crodafos" N-3 Neutral | 1.00 |
| "Fluilanol" | 1.00 |
| "Hartolan" | 1.00 |
| "Liquid base" | 0.50 |
| "Polawax" A-31 | 1.00 |
| "Skliro", lanolic acids | 1.00 |
| Stearic acid, triple pressed | 4.50 |

| | |
|---|---|
| Behenic acid | 1.50 |
| Lauric acid | 1.00 |
| 70% Sorbitol solution | 4.00 |
| Sodium alginate | 0.25 |
| Sequence B | |
| Diethanolamine | 2.00 |
| KOH to pH 9 | approximately 1.00 |
| Sequence C | |
| Distilled | 71.25 q.s. to 100% |
| Sequence D | |
| Potassium Sulfite | 9.00 |

Mix A and heat to 80°C. Dissolve B in a portion of C and add to sequence A at 80°C with agitation. When homogenous, add remaining water (C) at 80°C, continue agitation, then cool to 40°C and lastly add D. Continue agitation until salt is thoroughly dissolved.

Fill 80 parts by wt. into the can and 20 parts of a 7% $H_2O_2$ solution into the codispensing valve pouch, and cinch the valve to the can. Pressurize with 5% isobutane/propane (84:16) hydrocarbon mix. Most of the leading valve manufacturers have prototype codispensing valve units available for evaluation.

---

**Roll-on Antiperspirant**

| | |
|---|---|
| Phase I | |
| "Lantrol" AWS | 2.00 |
| Propylene glycol | 2.50 |
| S.D.A. 40, anhydrous | 20.00 |
| "Methocel" HG 0.65-4000 cps. | 0.75 |
| Phase II | |
| Water | 29.75 |
| Phase III | |
| Chlorhydrol 50% solution | 45.00 |

Slurry I with agitation and perfume. Slowly add II. Agitate slowly and Add III.

---

**Low Alcohol, Stick Anti-Perspirant**

| | |
|---|---|
| "Chloracel", 40% w/w Solution | 40.00 cc |
| Alcohol, SDA 40 | 25.00 cc |
| Propylene glycol | 23.00 cc |
| Na. stearate | 6.00 g. |
| Stearyl alcohol | 0.75 g. |

| | |
|---|---|
| Perfume | q.s. |

Mix "Chloracel", alcohol and propylene glycol and heat to 65 to 70°C. Dissolve the sodium stearate and stearyl alcohol in the mixture. After both have dissolved, the formulation is cooled and the perfume is added. The product is then poured into suitable containers.

---

**Aerosol Deodorant**

No. 1

(Concentrate)

| | |
|---|---|
| Heaxachlorophene | 0.5 |
| Zinc sulfocarbolate | 1.0 |
| "Acetol" | 1.0 |
| "Brij" 30 | 0.5 |
| SDA 40 anhydrous | 52.0 |

Mix all ingredients with agitation until in solution. Fill with either of the following propellant mixtures or blends of propellants between them to obtain desired spray characteristics.

| | A | B |
|---|---|---|
| Concentrate | 55.00 | 55.00 |
| "Freon" 114 | 19.35 | 27.00 |
| "Freon" 12 | 25.65 | 18.00 |
| | (colder) | (warmer) |

No. 2

| | |
|---|---|
| "Aerothene" MM solvent | 16.00 |
| "Aerothene" TT solvent | 9.50 |
| Ethanol SDA 40 | 38.75 |
| Hexachlorophene | 0.25 |
| Zinc phenosulfonate | 2.00 |
| Dipropylene glycol | 0.50 |
| Propellant 12 | 28.00 |
| Isobutane | 5.00 |
| Perfume | to suit |

---

**Facial Gel**

| | |
|---|---|
| Alcohol 39C | 50.0 |
| Allantoin | 0.2 |
| "Carbopol" 940 | 0.7 |
| Diisopropyl sebacate | 1.0 |
| Glycerin | 1.2 |
| Propylene glycol | 1.3 |
| "Dermodor" 6244 | 0.2 |
| "Tween" 20 | 0.4 |
| Water, distilled | 44.5 |
| Diisopropanolamine | 0.5 |

Add the carbopol to the alcohol

with fast stirring. Dissolve the allantoin into three-fourths part of the water and add to the alcohol. Add the diisopropyl sebacate and glycerin. Mix the perfume with the "Tween" 20 and propylene glycol and add to the mixture while stirring fast. Stop stirring to let the air bubbles rise to the surface. Dissolve the diisopropanolamine in the remaining quarter of the water with slight warming. While stirring very slowly, to prevent aeration, add the diisopropanolamin solution. Pack in tubes at once.

---

**Hydrophilic Ointment (Gel)**

| | |
|---|---|
| Stearyl alcohol | |
| White petrolatum | |
| Propylene glycol | |
| "Tween" | 20 |
| "Span" | 20 |
| Methylparaben | |
| Propylparaben | |
| "Avicel" R gel | (15%) |

The oil phase is composed of the stearyl alcohol and the white petrolatum. The water phase contains the "Avicel" R 15% gel, propylene glycol, emulsifying agent, and the parabens. Each phase is heated to 70°C on a water bath, and the water phase is added to the oil phase in a warmed mortar and pestle and triturated until congealed.

**Cream Mascara**

| | |
|---|---|
| Stearic acid | 75 |
| Beeswax | 85 |
| Carnauba wax | 20 |
| Morpholine | 33 |
| Color | 100 |
| Water | 650 |

Colors for mascara should be purchased from a cosmetic supply house or color dealer because no coal-tar dyes, and only a few earth colors, may be used.

Mix the first three ingredients. Stir gently, adding the color and heat to 185°F. Mix the morpholine and water, heat to 185°F, and add to first mixture slowly with stirring. Continue stirring and hold at 185°F until uniform. Set aside for several days before packaging.

---

**Cake Mascara**

| | |
|---|---|
| Stearic acid | 20 |
| "Aldo" 28 | 50 |
| Triethanolamine | 10 |
| Beeswax | 10 |
| Cosmetic pigment | 10 |

The ingredients are melted and stirred together. Grinding is usually not needed when finely ground cosmetic pigments are used. When carnauba wax is substituted for the beeswax, the triethanolamine and stearic acid should be increased so that the melting point of the mascara is not appreciably increased.

**Eye Make-up Remover**

| | |
|---|---|
| "Super-Sat" AWS 6 | 9 |
| "Super-Sat" AWS 3 | 21 |
| Mineral oil | 20 |
| P.E.G. 400 | 20 |
| Water | 40 |

**Face Powder, Polyethylene**

| | |
|---|---|
| Chalk, precipitated | 8 |
| Kaolin (cosmetic grade) | 6 |
| Polyethylene, micronized | 72 |
| Titanium dioxide | 6 |
| Zinc oxide | 2 |
| Zinc stearate | 6 |

---

**Military Face Camouflage Paint**

| | |
|---|---|
| "Castorwax" | 25.000 |
| Carnauba wax | 3.000 |
| Lanolin | 10.000 |
| Dimethyl phthalate | 20.000 |
| 2-Ethyl-1, 3-hexanediol | 10.000 |
| Color mixture (pigments only) | 32.000 |
| Phenyl mercuric benzoate (medicinal grade) | 0.004 |

Melt together and cast into stick form.

---

**Aerosol Talcum Powder**

wt./% of the entire aerosol

| | |
|---|---|
| Italian talc | 9.625 |
| "M.P."-51 | 0.375 |

The propellant is a 40/60 blend of propellants 12 and 114.

---

**Chapped-Lip Stick**

| | |
|---|---|
| Camphor | 20 |
| Beeswax, white | 18 |
| Petrolatum | 15 |
| Micro crystalline wax | 47 |
| Mineral oil | 62 |
| "Fluilan" | 60 |

Warm together gently; mix and pour. The above may be colored with a food dye and perfumed.

**Lipstick Base**

| | |
|---|---|
| Spermaceti | 92 |
| Vegetable oil | 40 |
| Diglycol laurate | 40 |
| White beeswax | 18 |

Add pigments as required.

Melt and mix at low temperature. Cool and pour into molds at lowest possible temperature to avoid settling of pigments. Hardness may be varied by proportions of beeswax or with addition of minute quantities of carnauba wax.

## Sun Screen Cream

No. 1

| | |
|---|---|
| Part A | |
| "Giv-Tan" F | 1.500 |
| Liquid petrolatum | 3.500 |
| White petrolatum | 15.000 |
| Stearic acid | 5.000 |
| Cety alcohol | 3.000 |
| "Polawax" | 2.000 |
| "Amerchol" L-101 | 6.000 |
| Part B | |
| "QUSO" F 22 | 5.000 |
| Part C | |
| Methylparaben | 0.025 |
| Propylparaben | 0.015 |
| Propylene glycol | 9.000 |
| Purified water | 50.000 |
| Part D | |
| Benzaldehyde | 0.055 |
| Rose centifolia | 0.066 |

Warm Part A to 75°C and place the sun screen agent in solutioh. Add the "QUSO" F 22 (Part B) to Part A with stirring until the silica has been incorporated.

Suspend the parabens in the propylene glycol and add to the purified water warmed to 75°C. Stir until the parabens are completely dissolved.

Slowly add Part C in small increments to the oil phase with stirring. Prior to solidification of the cream, add the perfumes (Part D).

No. 2

| | |
|---|---|
| Isobutyl-N-benzoyl-p-amino-benzoate | 4.0 |
| Isopropyl myristate | 40.0 |
| Glyceryl monostearate | 60.0 |
| Stearic acid | 8.0 |
| Propyleneglycol | 6.0 |
| Cetyl alcohol | 2.0 |
| Beeswax | 2.0 |
| Sorbic acid | 0.4 |
| Triethanolamine | 4.0 |
| Water | 128.0 |

Warm together and mix until uniform, then add color and perfume to suit.

**Sun Tan Oil**

| | |
|---|---|
| Propylene glycol monolaurate | 16.0 |
| Dipropylene glycol salicylate | 4.0 |
| "Robane" | 10.0 |
| Heavy mineral oil perfume and color q.s. | 100.0 |

**Sun Tan Lotion**

No. 1

| | |
|---|---|
| Lanolin | 80 |
| Stearic acid | 25 |
| Petrolatum | 25 |
| Triethanolamine | 8 |
| Preservative | 5 |
| Water | 775 |
| "Filtrosol" A | 70 |
| Perfume | |

Mix first three ingredients and heat at 194°F until melted. Mix water, preservative, and triethanolamine, heat to same temperature and add to first mixture with stirring. Continue stirring, and when cooling to 113°F has occurred, add "Filtrosol" A containing perfume. Stir until room temperature is reached.

No. 2

| | |
|---|---|
| Menthyl salicylate | 40 |
| Glycerol monostearate | 20 |
| Stearic acid | 15 |
| Cetyl alcohol | 5 |
| Triethanolamine | 10 |
| Benzyl alcohol | 6 |
| Water | 900 |
| Perfume | |

Mix all the ingredients except the water and heat to 175°F with stirring. When the mixture is clear, heat the water to 175°F and add it, continuing stirring until cold. Perfume before product solidifies.

No. 3

| | |
|---|---|
| Carboxymethyl cellulose | 100 |
| "Surfonic" N-40 | 5 |
| Polyethylene glycol 1450 | 50 |
| Bentonite | 30 |
| Magnesium oxide | 10 |
| Mineral black | 5 |
| Menthyl salicylate | 20 |
| Titanium dioxide | 130 |
| Petrolatum | 20 |
| Isopropyl alcohol | 200 |
| Water | 420 |
| Perfume | |

Dissolve the carboxymethyl cellulose in the water by vigorous stirring and add the other ingredients in the above order. Add perfume if desired.

No. 4

| | |
|---|---|
| "Filtrosol" A | 80 |
| Mineral oil | 60 |
| Stearic acid | 17 |
| Propylene glycol monostearate (Self-emulsifying) | 34 |

| | |
|---|---|
| Triethanolamine | 8 |
| Preservative | 10 |
| Water | 800 |
| Perfume | |

Mix first four ingredients and heat to 194°F. Dissolve triethanolamine in 400 ml of the water and heat to same temperature. Add it slowly to other mixture with gentle stirring. Add preservative to remainder of the water, heat to 194°F, and add to other mixture. When cooling to 104°F has occurred, add perfume and stir until cool.

No. 5

| | |
|---|---|
| Phase I | |
| Stearic acid | 2.50 |
| Isopropyl myristate | 4.00 |
| "Lantrol" | 3.00 |
| Silicone fluid 200—1000 cps. | 1.00 |
| Dehydroacetic acid | 0.10 |
| Dipropylene glycol salicylate | 2.00 |
| Homo menthyl salicylate | 0.80 |
| DLTDP | 0.05 |
| Phase II | |
| Water | 54.71 |
| Glycerin | 1.50 |
| Propylene glycol | 4.50 |
| 3% "Carbopol" 934 solution | 15.00 |
| "Sequestrene" Na 4 | 0.01 |
| "Tegosept" M | 0.18 |
| Phase III | |
| Water | 10.00 |
| KOH | 0.65 |

Heat all phases to 78°C. Add I to II with agitation. Add III. Agitate to 50°C and perfume. Cool down to 32°C with agitation and fill.

---

**Skin Browning, Aerosol**

| | |
|---|---|
| Erthrylose | 3.0 |
| Propyleneglycol | 0.2 |
| Alcohol, anhydrous | 96.8 |

Mix the above. To 33 parts add

| | |
|---|---|
| "Freon" 11 | 33.0 |
| "Freon" 12 | 33.0 |

**Skin Browning, Cream**

| | |
|---|---|
| Erythrylose | 4.0 |
| "Lanette" wax | 5.0 |
| Warm and mix then add warm water | 91.0 |

**Nail Polish Remover**

| | No. 1 | No. 2 |
|---|---|---|
| "Lantrol" | 0.5 | 0.5 |
| Ethyl acetate | 86.0 | 43.0 |
| Acetone | — | 43.0 |
| Ethyl alcohol | 10.0 | 10.0 |
| Water | 3.5 | 3.5 |

With agitation add "Lantrol" to ethyl acetate or acetone, add ethyl alcohol with agitation, and slowly with agitation add water.

**Dental Products**

(Toothpaste)

| | |
|---|---|
| Calcium carbonate | 20.0 |
| Calcium phosphate dibasic, hydrous | 32.0 |
| "QUSO" G 32 | 4.0 |
| Sodium lauryl sulfate | 1.2 |
| Sodium carboxymethylcellulose solution (0.5%) | 3.2 |
| Glycerin | 6.0 |
| Mineral oil | 1.0 |
| Sorbitol solution (70%) | 32.0 |
| Saccharin sodium | 0.6 |
| Peppermint oil | as needed |

Dissolve the saccharin sodium in the sodium carboxymethylcellulose solution and mix this solution together with the glycerin and sorbitol. Slowly add the "QUSO" with thorough mixing. Blend the dry powders, calcium carbonate, calcium phosphate and sodium lauryl sulfate together, and incorporate into the binding solution. Finally, add the mineral oil and peppermint oil.

**Germicidal Tooth Powder**

| | |
|---|---|
| Calcium carbonate, precipitated | 94.8 |
| Soap powdered | 5.0 |
| Cetyltrimethylammonium-cyclohexyl sulfamide | 0.1 |

| | |
|---|---|
| Peppermint oil | 0.1 |

---

**Mouth Wash**
("Lavoris" Type)

| | |
|---|---|
| Zinc chloride | 2.4 g. |
| Menthol | 0.8 g. |
| Soluble saccharin | 0.4 g. |
| Formaldehyde | 0.4 cc. |
| Clove oil | 0.4 cc. |
| Cinnamon oil | 1.6 cc. |
| Alcohol | 1.6 cc. |
| Distilled water to make | 1000.0 cc. |
| Liquid carmine, N.F. | 70.0 min. |

Dissolve the zinc chloride and the soluble saccharin in 800 cc of distilled water and add the formaldehyde solution. Dissolve the menthol and the oils in the alcohol and gradually add the aqueous solution to the alcoholic solution. Finally add sufficient distilled water to make 1000 cc. Allow the mixture to stand for 24 hr and filter until clear.

---

**Bactericidal Detergent**

| | | | |
|---|---|---|---|
| I | | Propylene glycol | 19.00 |
| | | Hexachlorophene | 3.00 |
| II | (1) | "Miranol" $C_2M$ | 18.10 |
| | (2) | "Sipon" LT 6 | 19.20 |
| III | (3) | "Lantrol" | 1.00 |
| | (3) | "Ethoxyol" 16q | 3.50 |
| | (4) | "Schercomid" SCO extra | 3.00 |
| IV | | Water | 18.20 |
| V | | Methocel mix | 5.00 |
| | | Water | 10.00 |

Methocel Mix:

| | |
|---|---|
| "Methocel" 65 HG 4000 | 3.00 |
| Propylene glycol | 10.00 |
| Water | 87.00 |

Heat I to 78°C. Add II, continue heating to 78°C. Heat III to 78°C. Add slowly to I and II with agitation. Heat IV to 78°C. Add to 1, II, and III. Agitate. Cool to 50°C. Perfume, agitate. Add V. Cool to 30°C.

### Anesthetic Lubricant

| | |
|---|---|
| "Intracaine" hydrochloride | 2.000 |
| Methylcellulose powder, 4000 cps | 1.000 |
| Butyl *para*-hydroxybenzoate | 0.015 |
| Methyl *para*-hydroxybenzoate | 0.130 |
| Distilled water | 100.000 |

Dissolve the preservatives in 75 cc of the water which has been heated to boiling. Use 25 cc of this solution while hot to wet the methylcellulose. Cool the remainder of the preservative solution before adding to the wet methylcellulose, shake, and place in a refrigerator until a clear solution of the methylcellulose is obtained. Add the Intracaine hydrochloride which has previously been dissolved in the remaining 25 cc of distilled water and mix thoroughly. Autoclave 15 min at 121°C in 30 cc wide-mouth, tightly capped bottles or in an Erlenmeyer flask covered with glassine paper and gauze. Pour into sterile jars.

---

### Surgical Lubricant

| | | |
|---|---|---|
| Sodium carboxymethylcellulose (CMC-70-High) | | 1.50 g |
| Propylene glycol, U.S.P. | | 25.00 cc |
| Methyl "Paraben", U.S.P. | | 0.15 g |
| Perfume | | 0.10 cc |
| Distilled water | qs ad | 100.00 cc |

Dissolve the methyl "paraben" in the propylene glycol, then add the sodium carboxymethylcellulose to this mixture and mix thoroughly. Heat the distilled water to boiling and gradually add the propylene glycol mixture to the boiling water with constant stirring. Stir until cool, then incorporate the perfume. Viscosity of the gel may be varied to suit needs for tubing or bottling.

---

### Surgical Scrub, Emollient

| | |
|---|---|
| Phase I | |
| "Triton" × 200 | 50.00 |
| Phase II | |
| "Supermide" L-9 | 2.00 |
| Phase III | |
| "Lantrol" | 5.00 |
| "Nimlesterol" D | 3.00 |
| Phase IV | |
| Propylene glycol | 5.00 |
| Hexachlorophene | 3.00 |

| | |
|---|---|
| Water | 32.00 |

Heat I, II, and III to 55°C. Heat IV until in solution. Add Phase II to Phase I. Add Phase III. Add Phase IV. Add water to mixture of I, II, III and IV. Make all additions with steady agitation.

---

**Emollient Skin Oil**

| | |
|---|---|
| "Lantrol" | 40.0 |
| Isopropyl myristate | 25.0 |
| Myristyl alcohol | 10.0 |
| Ethyl alcohol (95%) | 25.0 |
| Perfume | q.s. |

Heat all ingredients, except ethyl alcohol, until myristyl alcohol goes into solution. Stir down to cool and slowly add ethyl alcohol with continued stirring. Color and perfume.

---

**Improved Calamine Lotion**

| | |
|---|---|
| Zinc oxide | 8 g |
| Calamine | 8 g |
| Polyethylene glycol 400 | 8 cc |
| Polyethylene glycol 400 monostearate | 3 g |
| Lime water | 60 cc |
| Water, to make about | 100 cc |

Dilute 60 cc of lime water to 90 cc with water and heat to boiling. Heat 3 g of polyethylene glycol 400 monostearate in a small beaker to 70°C on a water bath. Stir the hot lime water solution on an electric stirrer while adding the melted monostearate slowly, continuing the the temperature of the mixture drops to about 40°C. Then set aside for about 24 hr. Mix 8 g of calamine and 8 g of zinc oxide in a mortar and prepare a smooth paste with 8 cc of polyethylene glycol 400; add the monostearate suspension slowly with continuous trituration. Transfer to a beaker and stir vigorously with an electric stirrer for 15 min.

*Note:* Care should be taken to incorporate as little air as possible during the stirring procedure.

---

**Hexachlorophene Sulfur Lotion**

| | |
|---|---|
| Hexachlorophene | 1.00 |
| Sulfur | 5.00 |
| Calamine | 3.00 |
| Cetyl alcohol | 1.50 |
| White wax | 1.00 |
| Sodium lauryl sulfate | 0.50 |
| Glycerin | 10.00 |
| Propylene glycol | 2.00 |
| "QUSO" F 22 | 2.00 |
| Methylparaben | 0.15 |
| Propylparaben | 0.05 |
| Purified water | q.s. |

Dissolve the parabens in forty percent of the final volume of purified water with the aid of heat. To the paraben solution add the sodium lauryl sulfate in half of the glycerin. Warm the combined solutions to 80°C. Melt the white wax and cetyl alcohol together and warm to 80°C. Add the aqueous phase slowly to the oil phase with constant stirring. Stir until the temperature of the emulsion reaches 40°C. Dissolve the hexachlorophene in the propylene glycol and add to the emulsion. Wet the sulfur and the calamine with the remaining glycerin and add to the emulsion with stirring. Add the "QUSO" F 22 with agitation and make the preparation up to volume with purified water. Pass the suspension through a colloid mill or homogenizer.

---

**Hospital Massage Lotion**

No. 1

| | |
|---|---|
| Olive oil | 560.0 |
| Stearic acid | 600.0 |
| Cetyl alcohol | 300.0 |
| "Aquaphor" | 375.0 |
| Aluminum dihydroxy allantoinate | 30.0 |

No. 2

| | |
|---|---|
| Triethanolamine | 250.0 |
| Sorbitol solution (70%) | 1.0 |
| Perfume | |
| Ethyl alcohol (95%) | 100.0 |
| Purified water | 11.4 |

Levigate allantoinate with olive oil. Heat Group 1 ingredients to 70°C. Place triethanolamine in water and heat to 70°C. Slowly add melted Group 1 ingredients to aqueous solution. Add sorbitol solution. Mix perfume with alcohol. Add alcoholic mixture to lotion when it has reached room temperature and stir well.

---

**Hydroalcoholic Rubbing Lotion**

| | |
|---|---|
| Part A | |
| "QUSO" F 22 | 5.000 |
| "Tween" 85 | 2.000 |
| Purified water | 53.000 |
| Part B | |
| Methylparaben | 0.200 |
| Menthal | 0.205 |

| | |
|---|---|
| Camphor | 0.025 |
| Ethanol (95%) | 39.750 |

Place the "QUSO" F 22 in a container and wet the silica with two-thirds volume of the water. Transfer to Waring Blender and agitate for 1 min. Dissolve the "Tween" 85 in the remaining water, add to the silica suspension, and agitate in the blender for 1 min. Dissolve the camphor, menthol and methylparaben in the ethanol and add to the aqueous silica suspension with stirring.

---

**Antiphlogistine Type Rub**

| | |
|---|---|
| Methyl salicylate | 12.0 |
| Eucalyptus oil | 0.5 |
| Menthol | 1.0 |
| Camphor | 1.0 |
| Vanishing cream to make | 100.0 |

---

**Iodized Emulsion**

| | |
|---|---|
| White wax | 5 |
| Cetyl alcohol | 5 |
| Heavy liquid petrotum | 15 |
| "Aerosol" OT | 3 |
| Water | 72 |
| Iodine | 1 |

The fats are melted together and heated to about 70°. The emulsifier is dissolved or mixed with the water and heated to the same temperature. The two solutions are mixed and stirred until cool. If necessary, a mechanical mixer is used to assist in obtaining a satisfactory emulsion.

---

**Medicinal Oil Emulsions**

| | No. 1 | No 2 | No. 3 |
|---|---|---|---|
| Heavy mineral oil | 13.55 | — | — |
| Cod-liver oil | — | 20.8 | — |
| Castor oil | — | — | 9.5 |
| Magnesium trisilicate | 8.25 | 12.0 | 10.8 |
| Water to make | 100.00 | 100.0 | 100.0 |

The water is added to the trisilicate with mixing. Then the oil is added to this gradually with high speed stirring.

---

**W/O Emulsion Type Base**

(For topical drug suspension)

| | |
|---|---|
| Liquid petrolaum | 45 |
| Stearyl alcohol | 10 |
| "QUSO" F 22 | 6 |
| Purified water | 39 |

Combine the stearyl alcohol with the mineral oil and warm on water bath until mixture is liquefied. Disperse the "QUSO" F 22 with agitation. Finally incorporate the purified water with thorough mixing.

**Drug Suspension Base, Topical**

| | |
|---|---|
| "QUSO" F 22 | 10 |
| Purified water | 5 |
| Isopropyl palmitate | 85 |

Disperse the "QUSO" F 22 in the isopropyl palmitate to form a translucent gel. Slowly incorporate the water.

---

**Sulfur Suspension**

| | |
|---|---|
| Sulfur, precipitated | 4.00 |
| "QUSO" F 22 | 5.00 |
| "Brij" 30 | 1.00 |
| Methylparaben | 0.15 |
| Propylparaben | 0.05 |

Dissolve the parabens in three-fourths volume of water by warming. Slowly add the "QUSO" while mechanically stirring the suspension. Then add the sulfur and disperse. After the sulfur has been dispersed, pass the suspension through the Manton-Gaulin colloid mill once, with the setting at 0.005 in. Disperse the "Brij" 30 in water and slowly add to suspension. Make the solution up to volume and stir for one minute.

---

**Methapyrilene Hydrochloride Suspension**

| | |
|---|---|
| Methapyrilene hydrochloride | 2.0 |
| Titanium dioxide | 5.0 |
| Benzocaine | 3.0 |
| "Brij" 30 | 1.5 |
| "QUSO" F 22 | 5.0 |
| Sorbic acid | 0.2 |
| Purified water | q.s. |

Place the "QUSO" F 22 in a container and wet the silica with two-thieds volume of the water. Add the titanium dioxide and benzocaine and transfer to a Waring Blender; agitate for one minute. Disperse the "Brij" 30 in purified water containing the sorbic acid and add to the suspension. Make the solution up to volume and agitate for one minute in the blender.

---

**Antacid-Demulcent Suspension**

| | |
|---|---|
| Aluminum hydroxide | 4.00 |
| Magnesium trisilicate | 4.00 |
| Carboxymethylcellulose | 1.00 |
| Methylparaben | 0.20 |
| Propylparaben | 0.04 |
| Sorbitol solution | 20.00 |
| Peppermint oil, terpeneless | 0.01 |
| Water, distilled, q.s. | 100.00 |

Dissolve the parabens in boiling water. Dissolve the CMC in the water. Allow gel to form. Disperse the aluminum hydroxide in the gel with agitation. Add some water. Allow the magnesium trisilicate to hydrate in the gel mixture; then disperse with agitation. Add the sorbitol and the flavor. Bring to full volume. Pass through a colloid mill and a straining tank.

**Calcium Carbonate-Magnesium Oxide Antacid Suspension**

| | |
|---|---|
| Calcium carbonate | 10.00 |
| Magnesium oxide gel | 6.70 |
| Methylcellulose 4000 cps. | 0.75 |
| Saccharin calcium | 0.02 |
| Methylparaben | 0.20 |
| Propylparaben | 0.04 |
| Peppermint oil | 0.01 |
| Water, distilled q.s. | 100.00 |

Dissolved the parabens in boiling water. Disperse the methylcellulose with agitation. Allow to cool and gel to form. Add water. Slurry magnesium oxide in gel mixture with agitation. Disperse calcium carbonate with agitation. Add saccharin, peppermint oil, and bring to full volume. Pass through colloid mill and a straining tank.

**Stomach Antacid**

| | |
|---|---|
| Magnesium hydroxide | 8.00 |
| "QUSO" G 32 | 5.00 |
| Sodium benzoate | 0.20 |
| Peppermint oil | 0.05 |
| Purified water | q.s. |

Dissolve the sodium benzoate in eight per cent of the final volume of water. Slowly add the "QUSO" G 32 while stirring, then the magnesium hydroxide. Make the suspension up to final volume, add the peppermint oil and pass through a colloid mill.

**"Digestive" Tablets**

| | |
|---|---|
| Heavy magnesium carbonate | 780 |
| Magnesium hydroxide | 29 |
| Light magnesium oxide | 34 |
| Calcium carbonate | 6716 |
| Kaolin | 34 |
| Calcium phosphate | 11 |
| Peppermint oil | 26 |

**Phenobarbital, 30 MG. Tablets**

| | Per Tablet |
|---|---|
| Phenobarbital | 30.0 mg. |

| | |
|---|---|
| Cellulose, microcrystalline | 30.0 mg. |
| Lactose, spray-dried | 70.0 mg. |
| "QUSO" F 22 | 1.3 mg. |
| Stearic acid | 1.3 mg. |

Screen phenobarbital chemical to remove lumps and blend with the cellulose. Add the lactose and blend. Finally add "QUSO" F 22 and stearic acid with thorough mixing. Compress; die, 9/32 in.; punch, shallow concave.

---

**Reserpine Tablet, 0.5 MG.**

| | Per Tablet |
|---|---|
| Reserpine | 0.50 mg. |
| Calcium phosphate dibasic, anhydrous | 75.00 mg. |
| Lactose, spray-dried | 75.00 mg. |
| "QUSO" F 22 | 0.75 mg. |
| Stearic acid | 1.50 mg. |

Blend the calcium phosphate and the lactose. Add reserpine, "QUSO" F 22, and stearic acid; mix thoroughly. Compress; die, 9/32 in.; punch, shallow concave.

---

**Salicylamide Tablet, 300 MG.**

| | Per Tablet |
|---|---|
| Salicylamide | 300.0 mg. |
| Starch paste, 10% | q.s. |
| Corn starch | 6.0 mg. |
| "QUSO" F 22 | 1.5 mg. |
| Calcium stearate | 3.0 mg. |

Add sufficient quantity of starch paste (10%) to the salicylamide to granulate and pass moist mass through No. 8 screen. Tray and dry the granulation overnight at 110°F. Pass the dried granulation through a No. 16 screen. Bolt the "QUSO" F 22 and calcium stearate onto the granulation and blend. Compress; die, 13/32 in.; punch, standard curvature.

---

**Thiamine Hydrochloride Tablet, 25 MG.**

| | Per Tablet |
|---|---|
| Thiamine hydrochloride, U.S.P. | 25.0 mg. |
| Lactose USP | 175.0 mg. |
| "QUSO" F 22 | 1.0 mg. |

| | |
|---|---|
| Calcium stearate | 2.0 mg. |

Blend the ingredients to give a uniform mixture. Compress; die, 9/32 in.; punch, shallow concave.

---

**Tripelennamine Tablet, 25 MG.**

| | Per Tablet |
|---|---|
| Tripelennamine hydrochloride | 25.00 mg. |
| Calcium phosphate dibasic | |
| Anhydrous | 95.00 mg. |
| Acacia, spray-dried | 30.00 mg. |
| "QUSO" F 22 | 0.75 mg. |
| Stearic acid | 1.50 mg. |

Blend the acacia and tripelennamine hydrochloride together, add the calcium phosphate, and mix thoroughly. Finally, add and blend the "QUSO" F 22 and the stearic acid. Compress; die, 9/32 in. Shallow concave.

---

**Triple Sulfa Suspension**

| | |
|---|---|
| Sulfamerazine, microcrystalline | 2.54 |
| Sulfadiazine, microcrystalline | 2.54 |
| Sulfamethazine, microcrystalline | 2.54 |
| Sodium citrate | 0.78 |
| "QUSO" F 22 | 6.00 |
| Methylparaben | 0.15 |
| Propylparaben | 0.02 |
| Citric acid (5%) | q.s.—pH 5.6 |
| Syrup (85%) | 90.60 |
| Purified water | q.s. |

Dissolve the parabens and sodium citrate in the syrup by warming. Add "QUSO" F 22 with agitation. Add the sulfonamides to the solution with stirring; adjust the pH to 5.6 with 5% citric acid solution and make up to volume with purified water. Pass the suspension through a colloid mill or homogenizer.

---

**Potassium Chloride Syrup**

| | |
|---|---|
| Potassium chloride | 10.00 |
| Citric acid monohydrate | 0.75 |
| Saccharin sodium | 0.07 |
| Sucrose | 30.00 |
| Sorbitol solution 70% w/w | 20.00 |

| | |
|---|---|
| FD & C Red No. 2 | 0.01 |
| Methylparaben | 0.10 |
| Propylparaben | 0.02 |
| Lime oil | 0.01 |
| Lemon oil | 0.01 |
| Imitation cherry flavor | 0.10 |
| Water, distilled q.s. | 100.00 |

Dissolve parabens in a sufficient amount of boiling water. Then dissolve the potassium chloride, citric acid, and saccharin. Then dissolve sucrose. Dissolve the FD & C Red No. 2. Add sorbitol and flavors. Bring to full volume. Add talc or another clarifying agent and filter solution.

---

**Castor Oil Oral Emulsion**
(36.0%)

| | |
|---|---|
| Castor oil, USP | 36.00 |
| Vanillin | 0.02 |
| Methyl salicylate | 0.13 |
| Saccharin calcium | 0.13 |
| Citric acid monohydrate | 0.05 |
| Sodium benzoate | 0.20 |
| Propyl gallate | 0.01 |
| Methylcellulose 4000 cps. | 0.75 |
| Water, distilled q.s. | 100.00 |

Slurry the methylcellulose in the castor oil. Add the propyl gallate and methyl salicylate. Dissolve the vanillin, saccharin, citric acid, and sodium benzoate in water. Add the aqueous soultion to the oil slurry with rapid agitation. Stir until viscous. Bring to full volume.

---

**Gelatine Pastilles**

| | |
|---|---|
| Gelatin | 50.00 |
| Simple syrup | 70.00 |
| Distilled water | 75.00 |
| Soluble saccharin | 0.50 |
| "Butoben" | 0.05 |
| Oil of lemon | 0.50 |

The convenient size and shape of the pastilles make them an ideal form for topical administration of antibiotics. The soothing effect of the gelatin is pleasing to the patients and the fresh lemon flavor appeals to children as well as adults.

---

**Expectorant**

| | |
|---|---|
| Terpin hydrate | 6.00 |
| Methyl salicylate | 0.15 |
| Anise oil | 0.15 |
| Alcohol | 20.00 |
| Propylene glycol | 45.00 |

| | |
|---|---|
| Glycerin | 20.00 |
| Sorbitol solution, USP q.s. | 100.00 |

Dissolve the terpin hydrate in the glycerin and propylene glycol at 50°. Mix the flavors with the alcohol and add to the terpin hydrate solution at 25°. Bring to volume with sorbitol solution.

---

**Kaolin Mixture With Pectin**

| | |
|---|---|
| Kaolin | 20.0000 |
| Pectin | 1.0000 |
| "QUSO" F 22 | 0.5000 |
| Benzoic acid | 0.2000 |
| Sodium saccharin | 0.1000 |
| Glycerin | 2% |
| Peppermint oil | 0.0075 |
| Purified water | q.s. |

Disperse the "QUSO" F 22 in half the volume of water, pass through the Manton-Gaulin colloid mill once with the setting at 0.005 in. Add the kaolin and mix. Triturate the pectin and sodium saccharin with the glycerin. Dissolve the benzoic acid in one third the volume of boiling water and add to the pectin mixture. When all the pectin is dissolved, add peppermint oil and the kaolin-"QUSO" mixtue. Mix thoroughly and make up to volume.

---

**Irritable Bladder Relief**

| | |
|---|---|
| Potassium citrate | 24.0 g |
| Tincture hyoscyamus | 30.0 cc |
| Paregoric | 30.0 cc |
| Elixir saw palmetto and Sandalwood ad | 120.00 |

In less severe cases the following prescription is useful:

| | |
|---|---|
| Tincture hyoscyamus | 30.0 cc |
| Potassium citrate | 30.0 g |
| Distilled water ad | 180.0 cc |

---

**Russian Blood Replacement**

| | |
|---|---|
| Sodium chloride | 7.0 |
| Potassium chloride | 0.1 |
| Calcium chloride | 0.1 |
| Purified casein | 30.0-40.0 |
| Sodium bicarbonate | 2.5 |
| Water | 1000.0 |

A highly valuable property of this solution is its ability to withstand high temperature. The colloidal infusion is poured into ampoules and sterilized under 1.5 atm. without deteriorating the protein fraction. This preparation does not produce anaphylaxis and does not cause other toxic symptoms.

---

**Topical Anesthetic and Antipruritic**

| | |
|---|---|
| "Tegone" A | 20.00 |
| "Tegamine" P-13 | 0.50 |

| | |
|---|---|
| "Tegosept" P | 0.10 |
| Water | 74.00 |
| Propylene glycol | 2.00 |
| Polyethylene glycol 1000 | 2.00 |
| Dibucaine citrate | 1.00 |
| Citric acid | 0.25 |
| "Tegosept" M | 0.15 |

**Antiobiotic, Anesthetic and Antiinflammatory**

| | | |
|---|---|---|
| 1 | "Tegin" 515 | 15.00 |
| 1 | "Tegamine" P-13 | 0.70 |
| 2 | Water | 72.01 |
| 2 | Phosphoric acid | 0.12 |
| 3 | Water | 10.00 |
| 3 | Neomycin sulfate | 0.50 |
| 3 | Dibucaine base | 0.50 |
| 3 | Phosphoric acid | 0.17 |
| 4 | Hydrocortisone acetate | 1.00 |

Heat 1 & 2 to 80°C, slowly and with good agitation add 2 to 1, cool with agitation until emulsion becomes smooth, slowly and with good agitation add 3, continue cooling with agitation, mix in 4.

**Eczema-Dermatitis Products**

| | |
|---|---|
| Prepared calamine | 40.0 |
| Zinc oxide | 40.0 |
| Glycerin | 10.0 |
| Solution of calcium hydroxide, to make | 500.0 |

When the inert powdered contents of this lotion is too drying and irritating, a mixture of this lotion and olive oil, equal parts, is very soothing.

A simple dusting powder for wet or dry lesions consists of:

| | |
|---|---|
| Boric acid | 8.0 |
| Zinc oxide | 40.0 |
| Purified talc | 72.0 |

A combination which is usually surprisingly antipruritic consists of:

| | |
|---|---|
| Menthol | 2.5 |
| Boric acid | 60.0 |
| Purified talc enough to make | 120.0 |

**Ointment**

| | |
|---|---|
| Resorcinol | 6 |
| White wax | 8 |
| Hydrous wool fat | 40 |
| Petrolatum, to make | 120 |

**Firm Zinc Glycerogelatin**

| | |
|---|---|
| Zinc oxide | 10 |
| Glycerinated gelatin | 30 |
| Glycerin | 25 |
| Dist. water | 35 |

**Acne Lotion, Peeling**

A keratolytic ointment recommended as part of a carefully developed routine for treating acne, particularly for adolescents. Unlike proprietary products, it is a flexible formulation in which the active ingredients can be adjusted to give an optimum peeling effect without inflammation.

| | |
|---|---|
| Resorcin | 2.4- 9.6 |
| Precipitated sulfur | 2.4-12.0 |
| Zinc oxide | 15.0 |
| Talc | 15.0 |
| Glycerin | 10.0 |

| | |
|---|---|
| Alcohol | 40.0 |
| Water | 40.0 |

---

**Ointment for Ringworm**

The following formula has been used quite successfully in the treatment of ringworm:

| | |
|---|---|
| Salicylic acid | 0.5 |
| Iodine | 0.5 |
| Sulfur, precipitated | 5.0 |
| Triethanolamine | 6.0 |
| Zinc undecylenate | 20.0 |
| Undecylenic acid | 5.0 |
| Washable base | 63.0 |

Dissolve the iodine in the triethanolamine. Triturate the powders with a small amount of the washable base and finally combine and mix all ingredients. Mill if necessary, using stainless or porcelain grinder plates.

---

**Crab Louse Ointment**

| | |
|---|---|
| $\gamma$-Benzenehexachloride | 1 |
| Stearyl alcohol | 3 |
| Beeswax | 8 |
| Petrolatum, white | 82 |

Warm and mix until uniform.

---

**Dermatitis Scab**

| | |
|---|---|
| Salicylic acid | 2 to 4% |
| Mineral oil or | |
| Olive oil or | |
| Vegetable oil | q.s. |

Apply directly to crusted areas, leave on overnight and wash out the following with shampoo.

**Scabicide**

No. 1

| | |
|---|---|
| Benzyl benzoate | 80 |
| Hydrous wool fat | 25 |
| Glyceryl monostearate | 12 |
| Methyl cellulose | 9 |
| Water | 165 |
| Triethanolamine | 5 |
| "Carbitol" | 5 |

Mix the methyl cellulose with half the water at boiling temperature; let stand for 20 min with occasional agitation; then add the remainder of the water at room temperature or lower. Mix the benzyl benzoate, hydrous wool fat, glyceryl monostearate, triethanolamine and carbitol and heat gently. When lukewarm, add to the cellulose mixture.

No. 2

Soft soap
Isopropyl alcohol
Benzyl benzoate, of each, equal parts

About 150 g. is required per patient. The entire body is rubbed with soft soap, special attention being given to parts usually infested. The patient then soaks in a warm bath (100°F.) for 10 min, rubbing himself during the bath. While still wet, the body is brushed all over with the lotion for 5 min, special attention again being given to affected parts. The patient then rests, allowing the body to dry, then continues the brushing another 5 min.

The body is next gently dried with a towel and the patient dresses in the same clothes worn before treatment. After 24 hr another bath is taken and clean clothes put on. Underclothes are washed and preferably boiled. Bedclothes are boiled or otherwise sterilized.

---

**Barber's Itch and Athlete's Foot Treatment**

A bad case of barber's itch was cured with the following treatment. The strength of the solution to employ is determined by the tenderness of the skin. Silver nitrate solutions above 1% tend to burn tender skin, form blisters, and leave a black stain. However, they are extremely effective in stubborn cases.

No. 1

| | |
|---|---|
| Silver nitrate | 0.75 |
| Water | 99.25 |

No. 2

| | |
|---|---|
| Silver nitrate | 1.00 |
| Water | 99.00 |

No. 3

| | |
|---|---|
| Silver nitrate | 2.00 |
| Water | 98.00 |

The user of any of these formulas is urged to make trial tests on infected-uninfected skin zones before applying the preparation to a large area. The 1% solution is effective in treating a slowly spreading skin infection of the athlete's foot type. Application must be just beyond the border of the infection.

---

**Athlete's Foot Treatment**

(Lyon's Solution)

| | |
|---|---|
| Sodium propionate | 2.0 |
| Sodium caprylate | 2.0 |
| Propionic acid | 3.0 |
| Undecylenic acid | 5.0 |
| Salicylic acid | 5.0 |
| Copper undecylenate | 5.0 |
| Dioctyl sodium sulfosuccinate | 1.0 |
| Distilled water | 20.0 |
| Isoproply alcohol 91% ad | 100.0 |

---

**Antibiotic Cream**

| | |
|---|---|
| Glyceryl monostearate SE | 15.00 |
| Glycerin | 6.00 |
| Spermaceti | 5.00 |
| Liquid petrolatum, heavy | 4.00 |
| "Butoben" | 0.10 |
| "Aerosol" OT | 0.02 |
| Distilled water, q.s. ad | 100.00 |

The antibiotic is dissolved in a small quantity of sterile saline and incorporated with a sufficient amount of the prepared base. Dermatologic antibiotic cream should be kept in the refrigerator to help preserve potency.

### Hydrocortisone Cream

| | |
|---|---|
| Hydrocortisone alcohol (or acetate) | 1.0 |
| Chlorocresol | 0.1 |
| "Cetomacrogol" emulsifying wax | 9.0 |
| White soft paraffin | 15.0 |
| Liquid paraffin | 6.0 |
| Purified water, to make | 100.0 |

### Diaper Rash Cream

| | | |
|---|---|---|
| (1) | "Tegacid" | 16.0 |
| | "Neobee" M.5 | 5.0 |
| | Cetyl alcohol | 1.0 |
| | "Amerchol" CAB | 5.0 |
| | "Solulan" 98 | 1.0 |
| (2) | Allantoin N-acetyl DL-methionine | 0.2 |
| | Propylene glycol | 5.0 |
| | Distilled water | 66.5 |
| (3) | "Dermodor" 4775 | 0.3 |

Heat 1 and 2 to 75°C and with rapid agitation and 2 to 1. Add 3 at cream point.

### Ointment Base

| | |
|---|---|
| Cholesterol, USP | 0.2 |
| "Waxolan" | 5.0 |
| Sorbitan sesquioleate | 3.0 |
| Petrolatum, USP white | 71.8 |
| Water | 20.0 |
| Perfume and preservative | q.s. |

### Methyl Salicylate Cream

| | |
|---|---|
| "Tegin" 515 | 4.00 |
| Methyl salicylate | 40.00 |
| Stearic acid (triple-pressed) | 2.00 |
| "Tegosept" P | 0.10 |
| Water | 47.75 |
| Propylene glycol | 5.00 |
| Triethanolamine | 1.00 |
| "Tegosept" M | 0.15 |

### Sulfadiazine Ointment No. 254-R

| | |
|---|---|
| "Tegin" | 5.00 |
| "Tegin" P | 5.00 |
| "Tegosept" P | 0.10 |
| Water | 59.75 |
| Glycerin | 25.00 |
| "Tegosept" M | 0.15 |
| Sulfadiazine | 5.00 |

### Guy's Crude Coal Tar Ointment

| | |
|---|---|
| Crude coal tar | 7 |
| Burow's solution | 17 |
| Hydrophilic petrolatum, USP | 33 |
| Lassar's paste | 43 |

To make 100

Take up the Burow's solution in the hydrophilic petrolatum, then add the other ingredients and mix thoroughly. Age at least 2 wks before using.

---

**White's Crude Coal Tar Ointment**

| | |
|---|---|
| Crude coal tar | 5 |
| Zinc oxide | 5 |
| Corn starch | 45 |
| Petrolatum | 45 |
| To make | 100 |

Mix the crude coal tar and zinc oxide and allow to stand for at least 24 hr. Mix the corn starch and petrolatum, then mix in the crude tar—oxide combination. Mill until smooth.

---

**Analgesic Balm Ointment**

| | |
|---|---|
| Menthol | 10.0 |
| Menthyl salicylate | 10.0 |
| White wax | 5.0 |
| Petrolatum | 10.0 |
| "Arlacel" 80 | 2.5 |
| Lanolin, anhydrous | 45.0 |
| Water distilled q.s. | 100.0 |

Fuse wax, petrolatum, lanolin, and "Arlacel" at 70°C. Add menthol and methyl salicylate. Heat water to 72°C and add to oils with agitation.

---

**Emollient Ointment**

| | |
|---|---|
| **Oil Phase:** | |
| "Amerchol" L-101 | 12.0 |
| "Modulan" | 9.0 |
| Liquid petrolatum, USP heavy | 20.0 |
| Microcrystalline wax, m.p. 170 | 12.0 |
| Sorbitan sesquioleate | 2.0 |
| **Water Phase:** | |
| Sorbitol solution, USP | 2.0 |
| Water | 43.0 |
| Perfume and preservative | q.s. |

Add the water phase at 75°C to the oil phase at 75°C with moderate stirring. Continue mixing while cooling to room temperature. Homogenize before filling.

---

**Hydrophilic Ointment (Gel)**

| | |
|---|---|
| Stearyl alcohol | |
| White petrolatum | |
| Propylene glycol | |
| "Tween" | 20 |
| "Span" | 20 |
| Methylparaben | |
| Propylparaben | |
| "Avicel" R gel | (15%) |

The oil phase is composed of the stearyl alcohol and the white petrolatum. The water phase contains the "Avicel" R 15% gel, propylene glycol, emulsifying agent, and the parabens. Each phase is heated to 70°C on a water bath, and the water phase is added to the oil phase in a warmed mortar and pestle and triturated until congealed.

---

**Methapyrilene Hydrochloride Ointment**

| | |
|---|---|
| Methapyrilene hydrochloride | 2 |
| Liquid petrolatum | 45 |
| Stearyl alcohol | 10 |
| "Quso" F 22 | 6 |
| Purified water | 37 |

Warm the liquid petrolatum and stearyl alcohol to give an homogeneous mixture. Disperse the "Quso" F 22 with agitation. Dissolve the methapyrilene hydrochloride in the water and and add to the oil phase with thorough mixing.

**Astringent Suppositories**

| | |
|---|---|
| Benzocaine | 2 |
| Ephedrine hydrochloride | 1/16 |
| Oxyquinoline sulfate | 1/4 |
| Bismuth subgallate | 1 |
| Balsam of Peru | 1 |
| Cocoa butter q.s. | 30 |

For one suppository.

---

**Hemorrhoidal Suppositories**

| | |
|---|---|
| Aluminum dihydroxy allantoinate | 0.2 |
| "Metycaine" | 1.0 |
| Zinc oxide | 10.0 |
| "MYRJ" 52 | 2.0 |
| Glyceryl monostearate S.E. | 5.0 |
| "Wecobee" SS q.s. ad | 100.0 |

Melt the "Wecobee", "MYRJ" 52 and glyceryl monostearate S.E. Triturate the powders, add part of the melted base and levigate. Pour the mixture into the melted base and mix thoroughly.

---

**Burn Cream**

No. 1

| | |
|---|---|
| "Tegone" A | 10.00 |
| Isopropyl myristate | 10.00 |
| Benzocaine | 3.00 |
| PEG 200 monolaurate | 2.00 |
| PEG 400 monolaurate | 2.00 |
| Stearic acid (triple-pressed) | 2.00 |
| Spermacet | 2.00 |
| Cetyl alcohol | 2.00 |
| Ottasept extra | 0.30 |
| Dilauryl thiodipropionate | 0.05 |
| Water | 60.25 |
| Propylene glycol | 5.00 |

| | |
|---|---|
| Triethanolamine | 1.00 |
| Allantoin | 0.20 |
| "Versene" 100 | 0.05 |
| "Tegosept" M | 0.15 |

No. 2

| | |
|---|---|
| "Tegacid" Special | 13.50 |
| "Tegone" A | 5.40 |
| Isopropyl myristate | 8.10 |
| Peach kernel oil | 8.00 |
| Benzocaine | 2.00 |
| Hexachlorophene | 0.25 |
| Dilauryl thiodipropionate | 0.05 |
| "Dow Corning 200", fluid | 1.00 |
| Water | 56.19 |
| Propylene glycol | 5.00 |
| "Versene" 100 | 0.03 |
| Sodium lauryl sulfate (30%) | 0.33 |
| "Tegosept" M | 0.15 |

---

**Burn Dressing**

| | |
|---|---|
| Petrolatum | 500 |
| Sorbitol monoöleate (oil soluble) | 20 |
| Sorbitol monoöleate (water soluble) | 3 |
| Sulfathiazole powder | 50 |
| Distilled water | 500 |

The petrolatum and the sorbitols are mixed and heat-sterilized. The sulfathiazole is sterilized and dissolved in the sterile distilled water. The solution is then emulsified with the petrolatum mixture, with aseptic precautions throughout the process. A quantity of 3-by-12-in. strips of fine mesh gauze are placed in a sterile porcelain tray, and on top of these is placed a quantity of the ointment. The tray is covered and heat treated below 210°F, just long enough for thorough permeation of the ointment into the gauze. The use of such impregnated gauze strips makes possible a more uniform dressing and facilitates its application.

---

**Burn Spray**

| | |
|---|---|
| Benzocaine | 1.5 |
| Benzethonium chloride | 0.1 |
| Hexachlorophene | 0.1 |
| Isopropyl alcohol | 7.0 |

| | |
|---|---|
| Dipropyleneglycel | 1.0 |
| Citric acid | 0.1 |
| Propellent | 90.0 |

**Prickly Heat Lotion**

No. 1

| | |
|---|---|
| Menthol | 1 |
| Glycerin | 1 |
| Salicylic acid | 4 |
| Alcohol (95%) to make | 100 |

No. 2

| | |
|---|---|
| Menthol | 2 |
| Camphor | 2 |
| Bismuth subnitrate | 10 |
| Zinc oxide | 10 |
| Alcohol | 120 |
| Lime water | 120 |

**Deodorization**

This is a true deodorant, in contrast to the odor-masking agents sometimes employed. Although the exact mechanism is not understood, effectiveness is believed to be due to an adsorption or chemical fixation of the obnoxious odor. Odors arising from decay and putrefaction are prevented by inhibition of putrefactive organisms. Mopping or spraying of 1000 ppm active solution of "Hyamine" 2389 will generally eliminate the odors of garbage pails, fish, onion, buses, trains, and airplanes may be eliminated without resorting to the use of deodorant blocks or odorous disinfectants which themselves may be offensive.

**Embalming Fluid**

| | |
|---|---|
| "Hyamine" 1622 (crystals) | 0.1 |
| Formalin | 2.1 |
| Phenol | 6.7 |
| Ethyleneglycol | 8.3 |
| Isopropanol | 33.3 |
| Water | 49.5 |

**Preservatives for Pharmaceutical**

| Product | "Paraben" Concentrations (%) | |
|---|---|---|
| | Methyl | Propyl |
| Hexachlorophene cream | 0.120 | 0.200 |
| Lubricating jelly | 0.150 | 0.050 |
| Creams and lotions (+0.02% butyl) | 0.150 | 0.050 |
| Hydrophilic ointment | 0.025 | 0.015 |
| Ointment bases (with bentonite) | 0.150 | 0.050 |
| Vitamin C solutions | 0.180 | 0.020 |
| Ephedrine eye-drops | 0.026 | 0.014 |
| General pharmaceuticals (which cannot be heated) | 0.070 | 0.030 |

CHAPTER VI

# DETERGENTS

**Dry Cleaning Compound**

| | |
|---|---|
| "Triton" X-100 | 12.5 |
| "Triton" X-45 | 5.0 |
| "Dytol" B-35 | 2.5 |
| Water | 2.5 |
| "Stoddard solvent" | 77.5 |

Dissolve the "Dytol" B-35 in the water. Add "Triton" X-100 and "Triton" X-45; then add "Stoddard solvent".

Use one part to 100 parts cleaning solvent.

---

**Dry Cleaning Charge Soap**

| | |
|---|---|
| "Emcol" P10-59 | 2 |
| "Emcol' 5138A | 1 |
| "Stoddard Solvent" | 1-3 |

---

**Laundry Detergent**

No. 1

This formulation exhibits moderate foam and low viscosity suitable for automatic injection. It leaves little residue in the measuring cup or reservoir. The formulation has been thoroughly field and laboratory tested, with excellent results. It is recommended at ½ cup for top-loading machines, and at ¼ cup for front-loaders.

| | |
|---|---|
| Water | 26.30 |
| Carboxy methyl cellulose | 0.50 |
| KOH (50% solution) | 2.20 |
| Potassium silicate (39.4% solution) | 12.70 |
| "Acrysol" ASE-108 | 6.50 |
| Tetrapotassium pyrophosphate (60% solution) | 41.70 |
| "Triton" X-100 | 10.00 |
| Optical brightener | 0.10 |
| Dye | 0.01 |
| Perfume | Trace |

No. 2

| | No. 1 | No. 2 |
|---|---|---|
| "Triton" X-114 | 5.00 | 5.00 |
| Sodium tripolyphosphate | 40.00 | 40.00 |
| Sodium sulfate | 30.00 | — |
| Borax | 14.00 | — |
| Sodium silicate | 10.00 | 15.00 |
| Soda ash | — | 39.00 |
| Carboxy methyl cellulose | 1.00 | 1.00 |
| Water-soluble optical bleach | 0.03 | 0.03 |

---

**Enzyme Laundry Detergents**

No. 1

(Nonionic-Anionic)

| | |
|---|---|
| Polyoxyethylated fatty alcohols ($C_{10}$-$C_{14}$) | 10.00 |
| Sodium alkylaryl sulfonate | 25.00 |
| Sodium tripolyphosphate | 43.45 |
| Tetrasodium phosphate | 12.00 |
| Sodium metasilicate | 5.00 |
| Sodium carboxymethylcellulose | 1.00 |
| Lauric diethanolamide | 3.00 |
| Nonionic optical brighteners | 0.05 |
| "Protease" S-10D | 0.50 |

No. 2

(Nonionic-Anionic)

| | |
|---|---|
| Alkylphenoxypoly (ethyleneoxy) ethanol | 10.00 |
| Sodium alkylaryl sulfonate | 5.00 |
| Sodium tripolyphosphate | 35.00 |
| Borax | 10.00 |
| Sodium metasilicate | 5.00 |
| Sodium carbonate | 33.45 |
| Sodium carboxymethylcellulose | 1.00 |
| Nonionic optical brightener | 0.05 |
| "Protease" S-10D | 0.50 |

No. 3

Anionic-Nonionic

| | |
|---|---|
| Sodium alkylaryl sulfonate | 10.00 |
| Polyoxyethylene tridecyl ether urea complex | 10.00 |

| | |
|---|---|
| Sodium tripolyphosphate | 35.00 |
| Sodium metasilicate | 8.00 |
| Borax | 12.00 |
| Sodium sulfate | 23.50 |
| Sodium carboxymethylcellulose | 1.00 |
| "Protease" S-10D | 0.50 |

No. 4
(Nonionic-Anionic)

| | |
|---|---|
| Dialkylphenoxypoly (ethyleneoxy) ethanol | 10.00 |
| Sodium alkylaryl sulfonate (40% active) | 25.00 |
| Sodium tripolyphosphate | 40.00 |
| Sodium metasilicate | 8.00 |
| Sodium sesquicarbonate | 15.50 |
| Sodium carboxymethylcellulose | 1.00 |
| "Protease" S-10D | 0.50 |

No. 5
(Cold-Water)

| | |
|---|---|
| Sodium alkylaryl sulfonate (40% active) | 44.00 |
| Alkylphenoxypoly (ethyleneoxy) ethanol | 3.00 |
| Sodium tripolyphosphate | 15.00 |
| Sodium sesquicarbonate | 34.50 |
| Trisodium nitrilotriacetate | 2.00 |
| Sodium carboxymethylcellulose | 1.00 |
| "Protease" S-10D | 0.50 |

No. 6
(Nonionic-Soap)

| | |
|---|---|
| Polyoxyethylene tridecyl ether | 13.00 |
| Sodium stearate | 2.00 |
| Sodium tripolyphosphate | 35.00 |
| Sodium metasilicate | 8.00 |
| Sodium sesquicarbonate | 40.45 |
| Sodium carboxymethylcellulose | 1.00 |
| Nonionic optical brightener | 0.05 |
| Protease S-10D | 0.50 |

**Enzyme Pre-Soak**

No. 1

(Nonionic)

| | |
|---|---|
| Alkylphenoxypoly (ethyleneoxy) ethanol | 2.0 |
| Sodium tripolyphosphate | 35.0 |
| Trisodium nitrilotriacctate | 1.0 |
| Sodium sesquicarbonate | 61.5 |
| "Protease" S-10D | 0.5 |

No. 2

(Anionic)

| | |
|---|---|
| Sodium alkylaryl sulfonate (40% active) | 10.0 |
| Sodium tripolyphosphate | 40.0 |
| Borax | 12.0 |
| Sodium metasilicate | 5.0 |
| Sodium sulfate | 32.5 |
| "Protease" S-10D | 0.5 |

No. 3

| | |
|---|---|
| "Alcalase" Novo | 5-10 |
| Sodiumtripolyphosphate | 300-700 |
| Buffer | 20-100 |
| C.M.C. (50%) | 0-20 |
| Sodium sulfate | 625 |
| Synthetic detergent (Nonionic) | 50-150 |

Use about 20 g of above to 1 gal water. Adjust pH to 8.7-9.3.

---

**Rinse Additive**

No. 1

| | |
|---|---|
| "Triton" CF-10 | 47.5 |
| "Triton" H-66 | 10.0 |
| Water | 42.5 |

Forms clear solution from 0 to 56°C.

No. 2

| % "Triton" CF-21 | % Water | |
|---|---|---|
| (a) 90-100 | 10-0 | |
| (b) 50 | 28 | 22% isopropanol |
| (c) 30 | 30 | 20% isopropanol and |

| | | |
|---|---|---|
| (d) 48 | 45 | 20% "Triton" X-45<br>7% "Sarkosyl" NL-30 |

---

**Biodegradable Rinse Aid**

| | A | B | C | D |
|---|---|---|---|---|
| "Triton" DF-16 | 30.00 | 30.00 | 30.00 | 30.00 |
| "Triton" X-45 | 10.00 | 10.00 | 10.00 | 20.00 |
| "Triton" DF-12 | 10.00 | 10.00 | 10.00 | — |
| "Hyamine" 3500 | 0.25 | 0.25 | 0.25 | — |
| "Sarkosyl" NL-30 | 0.90 | — | — | — |
| Isopropanol | — | 10.00 | — | 10.00 |
| Phosphoric acid | — | — | — | — |
| Water | 48.85 | 39.75 | 49.75 | 40.00 |

| | E | F | G | H |
|---|---|---|---|---|
| "Triton" DF-16 | 50.00 | 30.00 | 35.00 | 35.00 |
| "Triton" X-45 | — | 10.00 | 15.00 | 15.00 |
| "Triton" DF-12 | — | 10.00 | — | — |
| "Hyamine" 3500 | — | — | — | — |
| "Sarkosyl" NL-30 | — | — | — | — |
| Isopropanol | — | 10.00 | 10.00 | — |
| Phosphoric acid | — | — | 5.00 | 5.00 |
| Water | 50.00 | 40.00 | 35.00 | 45.00 |

---

**Aerosol Laundry Starch**

| | |
|---|---|
| "LE-463" 35% | 1.00 |
| Vegetable starch | 4.00 |
| Methyl "Parasept" | 0.17 |
| Propyl "Parasept" | 0.03 |
| "Tergitol" NPX surfactant | 0.10 |
| Distilled water | 94.70 |
| Perfume | q.s. |

The "Parasept" preservatives are added to prevent starch spoilage and "Tergitol" NPX serves to wet the fabric. If foaming occurs during preparation or filling of the starch, 0.5 parts of "SAG 470" silicone antifoam may be added.

Slurry starch and preservatives with one half of the water and heat slowly with moderate agitation to 85 to 90°C. Hold at this temperature for 15 min. Force-cool by adding remaining water and (if possible) use water jacket. After cooling, add "LE-463," "Tergitol" NPX, and perfume, with agitation.

### Laundry Starch, Liquid

| | |
|---|---|
| Water | 242.00 gal |
| "Amaizo" 715 pearl starch | 200.00 lb |
| Borax | 14.00 lb |
| "Carbowax" 6000 | 2.00 lb |
| Bleach (5% available chlorine) | 1.50 lb |
| Preservative (37% formaldehyde) | 4.00 lb |
| "Heliogen" Blue A supra paste dispersion | 6.00 lb |
| Antifoam ("Sag" 470) | 0.25 oz |

Add water to kettle and heat (121 gal of water or 18.60 in.) 6.5 gal per in.—calibration. Add borax, "Amaizo" 715 pearl starch, Antifoam and "Carbowax". Bring ingredients (Steps 1 and 2) to boil and hold for 20 min. Add cold water and bring to volume (242 gal or 37.25 in.). Add bleach, preservative, color, etc. Screen and bottle.

### Fine Fabric Detergent (Liquid)

May be used for hand-washing woolens and synthetics, such as sweaters and lingerie.

"Triton" X-114
"Sipon" ES
Water

Optical brightener, dye and perfume may be added as required by the formulator.

### Household Laundry Softener

No. 1

| | 4% solids | 6% solids | 8% solids |
|---|---|---|---|
| "Varisoft" 222 | 425 lb | 425 lb | 425 lb |
| Water | 885 gal | 585 gal | 430 gal |

Water at the temperature in the water main is suggested. Pump or pour "Varisoft" 222 (preferably above 60°F) into the water while stirring. Avoid whipping air into the mix and unnecessarily creating foam. Once a uniformly smooth dispersion is achieved (about 10 to 15 min of stirring should be adequate), the product may be packaged.

Finally, color and perfume can be added before bottling. Addition of a few ounces of Dow-Corning "Antifoam" A will effectively eliminate foaming during bottling, if this appears to be a problem. Bottling temperature for glass is of no special concern, but if plastic bottles are being filled, the mix temperature must be below 95°F to prevent bottle collapse after complete cooling.

No. 2

| | |
|---|---|
| "Onyx" BTC 2125 P40 | 4 |
| "Ammonyx" 2194 P40 | 20 |
| Filler | 76 |

The filler may be sodium bicarbonate, sodium sulfate, salt or urea. Recommended use level: One oz. of product for 10, lb load in the final rinse cycle (100 ppm active germicide and 500 ppm softener based on weight of the fabric in the rinsing solution). This is to combine germproofing and diaper-rash control with adequate softening properties.

**Fine Sweater Detergent**

| | |
|---|---|
| "Triton" X-114 | 10 |
| "Sipon" ES | 10 |
| Water | 80 |

Optical brightener, dye, and perfume may be added.

---

**Textile Detergent-Liquid**

| | |
|---|---|
| "Emcol" 5100 | 30 |
| Ethylene diamine Tetra-acetic acid sodium salt | 1 |
| Water | 69 |

---

**Laundry Bluing**

| | | |
|---|---|---|
| "Calcocid" blue 3B, 200% | | 0.25 oz |
| Softened water | q.s. | 1.00 gal |

Dissolve the color in a quart of warm water. Strain through unbleached muslin. Add 3 qt water and mix thoroughly.

---

**Rug and Upholstery Shampoos**

No. 1

| | |
|---|---|
| "Avirol" 103 | 15.00 |
| Coco diethanolamide (1:2) | 2.00 |
| Water, color, perfume, preservative | 83.00 |

No. 2

| | |
|---|---|
| "Avirol" 103 | 25.00 |
| Coco diethanolamide (1:2) | 3.00 |
| Water, color, 0erfume, preservative | 72.00 |

No. 3

| | |
|---|---|
| "Avirol" 103 | 35.00 |
| Coco diethanolamide (1:2) | 4.25 |
| Water, color, perfume, preservative | 60.75 |

In each of the three formulas, it is necessary to melt ingredients together to about 45°C. Some formulators may wish to add 0.5% of an optical brightener such as DBA—115 and 0.25% of "Santophen" 1 to give a superior type of product.

| | No. 4 | No. 5 |
|---|---|---|
| "Plurafac" C-17 | 5 | — |
| "Plurafac" A-38 | — | 20 |
| $Na_5P_3O_{10}$ | 2 | — |
| $Na_2SO_4$ | — | 20 |
| Isopropyl alcohol | 8 | — |
| $Na_2CO_3 \cdot NaHCO_3 \cdot 2H_2O$ | — | 60 |
| Water | 85 | — |

No. 6 (dry)

| | |
|---|---|
| "Emcol" 5138 | 3 |
| Perchlorethylene | 20 |
| Mineral spirits | 38 |
| Water | 39 |

Mineral spirits, perchlorethylene and "Emcol" 5138 are stirred together until homogeneous. To this add water slowly with stirring and continue until white homogeneous emulsion is formed.

---

**Lens Cleaner**

| | |
|---|---|
| Isopropyl alcohol | 85.0 |
| "Triton" X-100 | 0.1 |
| Water | 14.9 |

---

**Removing Stains From Synthetic Fiber Carpets**

1. On any fresh stain, remove all excess moisture with paper tissues by blotting. Do not rub the stained area, since this spreads the stain.
2. On any dried stain, when the surface of the stain is hard and crushed, break it up and remove the loose material by vacuuming.
3. With stains that are not identifiable, try detergent and water first, then standard cleaning solvents. If stains are not easily removed, seek the assistance of a professional carpet cleaner.
4. For best results, spot-clean as soon as stain is noticed.

A commercial spot remover will take out all of the stains listed below, and for best results this should be tried first. If spot remover is not available, however, the following stains can be removed with powdered detergent and warm water, using only the suds. Be sure to sponge up excess detergent to minimize resoiling. Rinse with sponge dampened in clean, clear water. Repeat until all suds have been removed. Blot with tissues.

| | | |
|---|---|---|
| Alcoholic drinks | Cough syrup | Mustard |
| Baked beans | Fats | Permanent ink |
| Ball point ink | French dressing | Plaster |
| Beet juice | Fruit juice | Puddings |
| Blood | Hair tonic | Root beer |

| | | |
|---|---|---|
| Butter | Ice cream | Sherbet |
| Candy | India ink | Soup |
| Catsup | Jelly | Sugar |
| Chocolate | Library paste | Tea |
| Clay | Mayonnaise | Tobacco juice |
| Coffee | Milk | Vegetable glue |
| Cola | Mucilage | Washable ink |
| Cooking oil | Mud | Wines |

With the stains listed below use a cleaning fluid such as acetone. Blot with tissues.

| | | |
|---|---|---|
| All-Purpose glue, lacquers | Furniture stain or polish | Oil paint |
| | | Permanent ink |
| Argyrol | Latex paint | Permanganate |
| Ball point ink | Linseed oil | Rouge |
| Carbon black | Lipstick | Rubber cement |
| Colored paper | Mustard | Salad oil |
| Crayon | Nail polish | Shoe dye |
| | Oil | Wheat paste |

Problem Stains

| | |
|---|---|
| Animal glue — | Detergent, blot. Ammonia, blot. Detergent, blot. Rinse, blot. |
| Chewing gum — | Perchloroethylene, blot. |
| Iron Rust — | Vacuum. 10% Hydrofluoric acid, blot. Rinse, blot. |
| Water colors — | Paint, oil and grease remover, perchloroethylene, blot. Detergent, blot. Stripper, blot. Rinse, blot. |
| Egg, gravy — | Detergent, blot. Ammonia, blot. Rinse blot. |

---

**Nicotine Finger-Stain Remover**

| | |
|---|---|
| Hydrochloric acid, pure | 4 |
| Glycerin | 100 |
| Rose water, triple | 900 |

To use the lotion, saturate a piece of cotton with it and rub gently over the stained area. Several applications may be necessary before desired results are obtained.

### Marble Stain Remover

"Most stains on marble will require the use of a poultice made with white cleansing tissue and whiting or powdered household cleaner mixed with a little water. The poultice should be soaked in this solution and kept from drying out while it is on the marble. It can be covered with a piece of glass or a sheet of plastic to keep the moisture from evaporating while the stain is being drawn out, which may take from 1 to 48 hr.

---

### Spot Remover

| | |
|---|---|
| Eriethanolamine sulphite | 15 |
| Glycol monobutyl ether | 20 |
| Ethoxylated propylene glycol | 8 |
| Water | 57 |

---

### Leatherette Cleaner

| | |
|---|---|
| "Maprofix" TLS-500 | 10.0 |
| "Onyx-ol" WW | 3.5 |
| "Neutronyx" 600 | 3.5 |
| IsopropyW alcohol | 5.0 |
| Water | 78.0 |

---

### General Household Cleaners

No. 1

| | |
|---|---|
| Sodium alkyl aryl sulfonate | 23 |
| "Triton" X-102 | 12 |
| Lauric diethanolamide | 3 |
| Water, dye, perfume, to make | 100 |

| | No. 2 | No. 3 | No. 4 |
|---|---|---|---|
| "Plurafac" C-17 | 6.0 | 8.0 | 6.0 |
| LAS (60% active slurry) | 25.0 | 25.0 | 20.0 |
| AES (60% active slurry) | — | 15.0 | 15.0 |
| Fatty amide | 2.5 | 4.0 | 2.0 |
| Urea | 5.0 | — | — |
| Isopropyl alcohol | — | — | 9.0 |
| Ethyl alcohol | — | 7.0 | — |
| Sodium toluene sulfonate | 8.0 | — | — |
| Opacifier | 3.0 | 3.0 | 3.0 |
| Water | 50.5 | 38.0 | 45.0 |

| | No. 5 | No. 6 |
|---|---|---|
| "Plurafac" C-17 | 8.0 | 10.0 |
| LAS (40% active flake) | 45.0 | 40.0 |
| AES (60% active) | 10.0 | — |
| Fatty amide | 2.0 | 2.0 |
| $Na_2CO_3$ and/or $Na_2SO_4$ | 35.0 | 48.0 |

| | No. 7 | No. 8 | No. 9 |
|---|---|---|---|
| "Triton" X-100 | 20.0 | 15.0 | 12.0 |
| "Triton" X-301 | 50.0 | — | — |
| Alkyl aryl sulfonate (active material) | — | 15.0 | 16.0 |
| Alkanolamide | — | — | 2.0 |
| Water, to make | 100.0 | 100.0 | 100.0 |

No. 10

| | | |
|---|---|---|
| 1. | "Neodol" 25-3 S | 2.00 |
| 2. | "Mona" ACO 100 | 0.50 |
| 3. | Sodium metasilicate (penta) | 2.00 |
| 4. | Trisodium phosphate | 1.00 |
| 5. | Butyl "Oxetol" | 3.50 |
| 6. | Steam-dist. pine oil | 0.25 |
| 7. | Softened water | 90.75 |

Treat warmed water, with 0.1% "Nullapon" BF 78 or similar chelating agent. Dissolve 3 and 4 with stirring. Melt 2 and add 1. Mix both and add to the solution. Add 5 and 6 while stirring. Color with uranine, quantity sufficient.

No. 11

| | |
|---|---|
| "Triton" QS-44 | 5.0 |
| Sodium hydroxide | 0.9 |
| "Triton" X-100 | 4.0 |
| TKPP | 16.0 |
| Water | 74.1 |

No. 12

| | |
|---|---|
| TKPP (anhydrous) | 10.0 |
| "Ivory" flakes | 5.0 |
| "Triton" X-102 | 5.0 |
| Propylene glycol | 3.0 |
| Water | 77.0 |

Warm and mix slowly.

No. 13

| | |
|---|---|
| Water | 76.8 |
| "Acrysol" ASE-108 | 3.0 |
| Sodium hydroxide | 0.2 |
| TKPP | 5.0 |
| TSPP | 5.0 |
| "Triton" X-100 | 8.0 |
| Lauric diethanolamide | 2.0 |

Preblend "Acrysol" ASE-108

with one part water. Add this mixture to remaining water and dissolve NaOH in small amount of water; add slowly with stirring. Add remaining ingredients in order listed, making sure that each is completely dissolved or dispersed before addition of the next.

No. 14

| | |
|---|---|
| "Klearfac" AA270 | 8.0 |
| "Plurafac" D-25 | 8.0 |
| Tetrapotassium pyrophosphate | 2.0 |
| Fatty amide | 2.0 |
| Deodorized kerosene | 5.0 |
| Monoethanolamine | 1.5 |
| Isopropanol | 20.0 |
| Pine oil | 2.0 |
| Water | 51.5 |

No. 15

| | |
|---|---|
| Plurafac" C-17 | 2.0 |
| Aqueous ammonia (30%) | 10.0 |
| Water | 88.0 |

## Pearly Liquid Detergent

| | No. 16 | No. 17 |
|---|---|---|
| "Triton" X-102 | 10 | 12 |
| Sodium alkyl aryl sulfonate (active) | 23 | 23 |
| Lauric diethanolamide | 5 | 3 |
| Ethyl alcohol | 2 | 2 |
| Ethyleneglycol monostearate | 2 | 2 |
| Water, to make | 2 | 2 |

In one method, 2% by weight ethylene glycol monostearate is added to the finished detergent blend at 75°C and the temperature held at this level until all of the crystalline material is melted. The heat is removed and the liquid allowed to cool to room temperature with constant, slow agitation. Pearlesence developed as the system cools.

An alternate method of preparation is by formulating a concentrated pearled system using all of the pearling agent in a portion of the "Triton" X-102 and water. This mixture has the following composition by weight.

| | |
|---|---|
| Ethylene glycol monostearate No. 40 | 15 |
| "Triton" X-102 | 15 |
| Water | 70 |

No. 18

| | |
|---|---|
| "Alipal" EO526 (57%) | 10 |
| NaOH (50%) | 3 |
| Tall oil fatty acid | 10 |

| | |
|---|---|
| "Hampshire" $NTANa_3$ Crystals | 10 |
| Water | 67 |

This type cleaner uses its alkalinity and organic surfactant for detergency, foam and "no-rinse" features. Prepare by adding the components in the order given with agitation until dissolved. No heating is required. The resultant product is a light-yellow, clear liquid with a viscosity of about 300 cps. The pH of approximately 10 is not high enough to damage floor tiles, painted surfaces, or rugs. Use concentrations range from 1 to 6 oz/gal. Product will not require rinsing since it foams in the bucket but not on the floor.

No. 19

(For hand make)

| | |
|---|---|
| "Alipal" CD-128 | 1.70 |
| Coconut acid-diethanolamine condensate | 0.50 |
| Sodium metasilicate | 1.70 |
| Tetrasodium pyrophosphate | 1.00 |
| Ethylene glycol monobutyl ether | 3.50 |
| Color, Perfume, water | balance |

No. 20

(Concentrated)

| | |
|---|---|
| Tetrapotassium pyrophosphate (60%) | 33.3 |
| "Alipal" CD-128 | 8.3 |
| Ammonium hydroxide (30%) | 3.4 |
| "Lytron" 611 | 0.1 |
| Color, perfume, water | balance |

No. 21

(Heavy Duty)

| | |
|---|---|
| House detergent, heavy duty (liquid) | 12.0 |
| "Petro" 11 powder | 9.0 |
| Linear alkyl sulfonate (100%) | 14.5 |
| Hexametaphosphate (63% $P_2O_5$) | 45.0 |
| Tripolyphosphate | 1.0 |
| Soda ash | 3.0 |
| Nonionic water | balance |

No. 22

General Purpose Cleaner: The following formulation makes an

excellent viscous, non rusting cleaner concentrate for use on floors, walls, equipment, etc.:

| | |
|---|---|
| "Varion" 1084 | 7.0 |
| Tetrapotassium pyrophosphate | 10.0 |
| Sodium xylene sulfonate (40%) | 4.0 |
| Water | 79.0 |

The xylene sulfonate is present, not as a coupler, but to modify viscosity. Without it, gels are obtained at this concentration.

No. 23
(High Viscosity)

| | |
|---|---|
| "Varamide" A10 | 5—10% |
| Tetrapotassium pyrophosphate | 4—6% |
| Sodium xylene sulfonate | 1—2% |
| Water | balance |

**Modern Bar Soap**

| | |
|---|---|
| "Neo-fat" 265 | 12.17 |
| "Neo-fat" 65 | 48.68 |
| 50% Sodium hydroxide (50%) | 18.72 |
| Water | 20.43 |
| Preservative | q.s. |

Add the fatty acids and about half the water to the soap crutcher heating until the fatty acids are melted (about 160°F). Add the rest of the water to the sodium hydroxide solution. Add the sodium hydroxide solution to the crutcher while the crutcher screw is turning. Test the finished soap for alkalinity or acidity by titrating in a neutral alcohol solution with 0.5N sodium hydroxide of 0.5N hydrochloric acid. The soap should finish between 0.1% free caustic and 1.0% free fatty acids (as oleic). Adjust with sodium hydroxide or fatty acids if necessary. Add the preservative. The finished soap is dried in a conventional dryer to a moisture content of 12-13%. Add perfume and 2-6% amine oxide (100% ·basis) to the soap chips and mill in a soap mill. This is followed by plodding and pressing.

**Detergent Cake**

No. 1

| | |
|---|---|
| Disodium $\alpha$ sulphostearate | 7 |
| Disodium $\alpha$ sulpho acids from fully hydrogerated coconut oil fatty acids | 7 |
| Sodium fatty acyl isothionate from coconut fatty | |

| | |
|---|---|
| acids | 32 |
| Sodium soap | 22 |
| Coconut oil fatty acid ethanolamide | 18 |
| Stearic acid | 3 |
| Imort | 5 |
| Water | 6 |

No. 2

To make a 12-g tablet, 30 mm in diameter and 13.1 mm in thickness. A crush strength of at least 17 lb is attainable. One tablet is equal to a tablespoon of granular material so 1 to 2 tablets should be used per load.

| | |
|---|---|
| Sodium metasilicate | 30.00 |
| Nonionic surfactant | 0.28 |
| Soda ash | 13.00 |
| White mineral oil | 1.67 |
| Sodium tripolyphosphate | 50.00 |
| Sodium dichloroisocyanurate | 4.00 |
| "Sterotex" | 0.50 |

## Liquid Hand Soap

| | | High-foam | Good-cleaning |
|---|---|---|---|
| A | "Neo-Fat" 265 | 15.00 | 5.00 |
| | "Neo-Fat" 94-04 | 15.00 | 21.00 |
| | Isopropanol | 6.00 | 12.00 |
| | Glycerin | 2.00 | 2.00 |
| | Water | 48.32 | 48.94 |
| B | 50% Potassium hydroxide | 13.68 | 11.06 |
| C | Perfume & preservatives | q.s. | q.s. |

Place A in a kettle and mix. When dispersion is complete add B and mix. Add C.

## Waterless Hand Cleaner

| | |
|---|---|
| "Triton" X-100 | 10.3 |
| Deodorized kerosene | 51.4 |
| Oleic acid | 4.1 |
| Stearic acid | 2.9 |
| Sodium hydroxide | 0.4 |
| Water | 30.9 |

The "Triton" X-100, stearic acid, and oleic acid are added to the kerosene and agitated until homogeneous. It may be necessary to predissolve the stearic acid in a portion of warm kerosene for ease of solution A caustic concentrate is prepared with a protion of the water and added to the mixture. Subsurface agitation is continued until the mixture is uniform. The balance of the water is then added slowly with continuous agitation until the mixture is smooth. Emollients, perfume, etc. may be added as desired. This formula will produce a white, opaque, heavy paste. Its viscosity may be increased by replacing all or part of the deodorized kerosene with a light mineral oil.

Mix "Triton" X-155 and orthodichlorobenzene well; then add 2/3 of the water and mix thoroughly. The hydrochloric (muriatic) acid is then added slowly and mixed well. The "Rhoplex" X-52 is diluted with the remaining 1/3 of the water. The final step is the addition of the diluted "Rhoplex" solution with stirring until the mix assumes a uniform milky-white appearance. No heat is necessary in preparing this product. This formulation yields a product containing 18% hydrochloric acid. If the acid concentration is increased above the recommended level, the stability of the formulation is destroyed and discoloration will develop.

This acid bowl cleaner should not be packaged in polyethylene, for the orthodichlorobenzene will bleed through the container. Glass is recommended for packaging.

No. 2

| | |
|---|---|
| "Atreol" | 60 |
| "Utrasene" | 25 |
| "Triton" X-207 | 15 |

No. 3

| | |
|---|---|
| "Tegin" 515 | 5.00 |
| Mineral oil (light N.F.) | 20.00 |
| Kerosene (deodorized) | 20.00 |
| Stearic and (triple pressed) | 3.00 |
| "Tegosept" P | 0.10 |
| Water | 45.25 |
| Propylene glycol | 5.00 |
| Triethanolamine, | 1.50 |
| "Tegosept" M | 0.15 |

No. 4

A. *Oil phase*

| | |
|---|---|
| Deodorized kerosene | 39.0 |
| Pale oleic acid | 7.5 |
| "Neodol" 25-3 | 2.0 |

B. *Water phase*

| | |
|---|---|
| Monoethanolamine | 0.8 |
| Triethanolamine | 2.6 |
| Propylene glycol | 2.5 |
| Glycerin | 1.0 |
| Lanolin | 0.5 |
| Butyl "Oxitol" | 0.4 |
| Water | 43.7 |
| Perfume, color | as desired |

Heat phases A and B separately to 70-75°C. Pour A into B while stirring continuously. Continue until a homogeneous smooth gel is formed, pour into jars. Let cool to room temperature.

No. 5

| | |
|---|---|
| "Emcol" 5138 | 3.0 |
| Deodorized kerosene | 49.0 |
| Water | 48.0 |

No. 6

| | |
|---|---|
| "Emcol" 5130 | 18.0 |
| Deodorized kerosene | 51.0 |
| Water | 31.0 |

Add "Emcol" 5130 to kerosene. Blend in water. $TiO_2$ may be added to whiten.

---

**Machine Dish Cleaners**

No. 1

| | |
|---|---|
| Sodium tripolyphosphate | 50.0 |
| Sodium metasilicate (anhydrous) | 47.0 |
| "Triton" CF-32 | 3.0 |

The addition of 2% sodium aluminate significantly reduces china over-glaze attack.

No. 2

| | |
|---|---|
| "Triton" CF-54 | 2.0 |
| Soda ash | 41.0 |
| Sodium tripolyphosphate | 30.0 |
| Sodium metasilicate (anhydrous) | 25.0 |
| Sodium dichloroisocyanurate (granular) | 2.0 |

Thoroughly blend the "Triton" CF-54 on the soda ash, sodium

tripolyphosphate and sodium metasilicate. Then add the granular sodium dichloroisocyanurate.

| | | Soft water formulation | |
|---|---|---|---|
| | | No. 3 | No. 4 |
| Sodium silicate ($SiO_2$:$Na_2O$ 2.00:1)[2] | | 13.0 | 13.0 |
| Defoaming nonionic surfactant | | 1.5 | 1.5 |
| Filler | | 46.8 | 48.0 |
| Specially prepared solution of: | | | |
| Sodium hydroxide | 0.1 | | — |
| STPP anhydrous granular | 0.1 | | — |
| Water | 4.5 | 5.0 | |
| Sodium aluminate | 0.3 | | — |
| STPP anhydrous granular | | 31.2 | 25.0 |
| STPP hexahydrate granular | | — | 8.0 |
| Sodium aluminate | | — | 2.0 |
| Sodium dichloroiscyanurate | | 2.5 | 2.5 |

| | Hard water formulation | | |
|---|---|---|---|
| | No. 5 | | No. 6 |
| Sodium silicate ($SiO_2$:$Na_2O$ 2.00:1)[2] | 17.0 | | 17.0 |
| Defoaming nonionic surfactant | 1.5 | | 1.5 |
| Filler | 19.6 | | 17.0 |
| Specially prepared solution of: | | | |
| Sodium hydroxide | | 0.1 | |
| STPP anhudrous granular | | 0.1 | 5.0 |
| Water | | 4.5 | |
| Sodium aluminate | | 0.3 | |
| STPP anhydrous granular | 54.4 | | 48.0 |
| STPP hexahydrate granular | — | | 12.0 |
| Sodium aluminate | — | | 2.0 |
| Sodium dichloroiscyanurate | 2.5 | | 2.5 |

| | No. 7 | No. 8 |
|---|---|---|
| "Plurafac" RA-40 or RA-43 | 3 | 2 |
| $Na_4P_2O_7$ | 35 | — |
| $Na_5P_3O_{10}$ | 15 | 61 |
| $Na_2CO_3$ | — | 15 |
| $Na_2SiO_3 \cdot 5H_2O$ | 10 | — |
| $Na_2SiO_3$ | — | 10 |

| | | |
|---|---|---|
| $Na_3PO_4$ (chlorinated) | 20 | — |
| $Na_2SO_4$ | — | 12 |
| Water | 17 | — |

**Enzyme Dish-Washing Powder**

| | |
|---|---|
| Sod. tripolyphosphate | 30.0 |
| Sod. hexametaphosphate | 30.0 |
| Sod. metasilicate | 10.0 |
| Sod. bicarbonate | 16.0 |
| Trisodium phosphate | 10.0 |
| "Pluronic" L62 | 3.0 |
| "Alcalase" | 0.5 |
| Bacterial amylase "Novo" 5000 SKB | 0.5 |

**Restaurant-Ware Rinse**

| | |
|---|---|
| "Triton" CF-10 | 50 |
| Isopropyl alcohol | 22 |
| Water | 28 |

The isopropyl alcohol serves as a solubilizing agent in the above formulation. This concentrate would be used by the consumer at the rate of 1 oz per 32 gal of rinse water (100 ppm active detergent).

**Dish Cleaner (Hand Type)**

| | No. 1 | No. 2 | No.3 |
|---|---|---|---|
| "Naxonate" 4L | 12.5 | 15.0 | 15.00 |
| ABS, sodium salt-57.5% act. | 28.0 | 25.0 | 35.00 |
| Urea | | | 2.00 |
| Alkanolamide | | 4.0 | |
| "Alipal" CO-436 (GAF) | 16.6 | 17.0 | 20.00 |
| "Emulphogene" BC-840 (GAF) | 4.0 | | 3.75 |
| Epsom salts | 1.0 | | 1.00 |
| E-284 Latex | 2.0 | 2.0 | |
| Phosphoric acid 75% | | to pH 7.2 | |
| Water | balance | balance | balance |

| | Opacified | Neutral, clear | Mildly alkaline, clear |
|---|---|---|---|
| | No. 4 | No. 5 | No. 6 |
| E-300-DL 40% | 0.5-2.0 | | |
| Sodium xylenesulfonate (40%) | 20.0 | 20.0 | 15.0 |

| | | | |
|---|---|---|---|
| "Tergitol" anionic 15-S-3A (60%) | 23.3 | 5.0 | 10.0 |
| "Tergitol" nonionic 15-S-12 (100%) | 6.5 | 5.0 | 3.0 |
| "Ucane" 11 (100% LAS) (sulfonate, sodium salt) | 19.1 | 20.0 | 20.0 |
| Acid (dilute sulfuric) | to pH 6.5-7.5 | to pH 6.5-7.5 | |
| $NTANa_3$ Crystals | 1.0 | 1.0 | 5.0 |
| KOH (50%) | — | — | 0.6 |
| Perfume, dye | | trace | |
| Water | | to 20%—40% solids | |

Disperse E-300-DL in 10 parts of water. To the balance of the water add the other ingredients and adjust pH. Finally add the diluted E-300-DL with good agitation.

No. 7

| | |
|---|---|
| "Ninex" 21 | 33 |
| Softened water | 67 |

Mix to uniform solution. Color and perfume optional.

No. 8 (Pink Milky)

| | |
|---|---|
| Sodium alkyl aryl sulfonate | 23.0000 |
| "Triton" X-102 | 12.0000 |
| Lauric diethanolamide | 3.0000 |
| Latex E-284 (40%) | 2.0000 |
| "Calcozine" rhodamine BX conc. | 0.0025 |
| Ethanol | 2.0000 |
| Water | 100.0000 |

For ease of preparation, the required amount of latex E-284 may be mixed with twice that volume of water, plus the calculated amount of alcohol. This blend is then added to the balance of the prepared formula. The introduction of the dye is most easily handled if a 0.05%-water solution is prepared as a stock mix. Experience indicates the color disperses readily and uniformly with no difficulty.

---

**Restaurant Rinse Additive**

| % "Triton" CF-10 | % Water | % Other additives | Cloud point (°F) |
|---|---|---|---|
| (a) 70-100 | 30-0 | None | 140 |

| | | | | |
|---|---|---|---|---|
| (b) | 50 | 30 | 20 Isopropanol | 140 |
| (c) | 50 | 40 | 10 Hydroxyacetic acid (active) | 126 |
| (d) | 30 | 32 | 20 "Triton" X-45 & 18 Isopropanol | 140 |
| (e) | 24 | 47 | 24 "Triton" CF-21 & 5 "Sarkosyl" NL-30 | 140 |

---

### Beverage Glass Cleaners

In the field of sanitary supply cleaners there is always a demand for dilute detergents with high viscosity. An economical and effective cleaner of this type can be made as follows, using a sodium alkyl aryl sulfanate as the base:

| | |
|---|---|
| "Stepan" DS-60 | 3.0 |
| "Ninol" AA62 Extra | 2.0 |
| Water | 95.0 |

A still cheaper formulation is the following:

| | |
|---|---|
| "Stepan" DS-60 | 2.0 |
| "Ninol" 128 Extra | 3.0 |
| Sodium tripolyphosphate | 0.5 |
| Water | 94.5 |

---

### Glass Window Cleaner

| | |
|---|---|
| "Genapol LRO" liquid | 1.5 |
| Isopropyl alcohol | 35.0 |
| Distilled water | 64.5 |

---

### Glass Cleaner, Liquid

| | |
|---|---|
| "Emcol" 5100 | 1.0 |
| Hexalene glycol | 3.0 |
| Isopropanol | 22.0 |
| Water | 74.0 |

### Bottle Washing Solution

| | |
|---|---|
| "Chelig"-32 | 1.2 |
| NaOH tetrasodium pyrophosphate | 12.0 |
| Water to 200 parts | 1.2 |

---

### Powdered Bottle Washing Compound

| | |
|---|---|
| "Triton" QS-44 | 1.25 |
| Organic sequestering agents | 1.00 |
| Sodium hydroxide | 97.75 |

---

### Floor Cleaner

No. 1

| | |
|---|---|
| "Triton" X-100 | 10.00 |
| Sodium hydroxide (10%) | 2.80 |
| "Acrysol" ASE-108 | 4.25 |
| Sodium nitrite | 0.15 |
| "Dowicide" G | 0.05 |
| Water | 82.75 |

Preblend "Acrysol" ASE-108 with 1 part water and add to remaining water. Add sodium hydroxide slowly with agitation. Then add sodium nitrite, "Dowicide" G, and "Triton" X-100 in that order, taking care to thoroughly blend each ingredient before adding the next one.

White vinyl floor tile may be

discolored by misuse of cleaners containing phosphate and nonionic detergent. Avoid excessive concentration and exposure time.

No. 2
(Low-Foam, Machine Scrubber)

| | |
|---|---|
| "Triton" X-114 | 3.00 |
| "Triton" H-66 | 4.00 |
| TKPP | 8.00 |
| Water, to make | balance |

No. 3
(Neutral)

| | |
|---|---|
| "Triton" X-100 | 20.00 |
| CMC 70 medium | 1.00 |
| "Dowicide" G | 0.05 |
| Sodium nitrite | 0.15 |
| Isopropylamine | 0.15 |
| Water | 78.65 |

Thoroughly dissolve CMC in the water, then add the "Triton" X-100 and other ingredients.

---

**"Bucket" Soap, Liquid**

| | |
|---|---|
| "Alipal" EO526 (57%) | 10 |
| NaOH (50%) | 3 |
| Tall oil fatty acid | 10 |
| $NTANa_3$ crystals | 10 |
| Water | 67 |

The resultant product is a light yellow, clear liquid with a viscosity of about 300 cps. The pH of approximately 10 is not high enough to damage floor tiles, painted surfaces, or rugs. Use concentrations range from 1 to 6 oz/gal. Product will not require rinsing since it foams in the bucket but not on the floor.

---

**Heavy-Duty Cleaners**

| | No. 1 | No. 2 |
|---|---|---|
| | % on 100% Active Basis | |
| "Naxonate" 4L | 5.0 | 4.0 |
| ABS, acid | 20.0 | 9.0 |
| Diethanolamine | 7.2 | 3.3 |
| "Igepal" CO-630 (GAF) | | 3.0 |
| Tetrapotassium pyro phosphate | 12.0 | 10.0 |
| Potassium silicate | 3.0 | 4.0 |
| CMC (100%) | 1.0 | 1.0 |
| Optical brightner | 0.1 | 0.1 |
| Water | balance | balance |

| | No. 4 | No. 5 | No. 6 |
|---|---|---|---|
| "Plurafac" A-38 | — | — | 12 |
| "Plurafac" B-26 or D-25 | 8 | 10 | — |
| "Plurafac" B-26 (optional duduster) | — | — | 1 |
| "Carbose" D (CMC) | 1 | 2 | 2 |

| | | | |
|---|---|---|---|
| $Na_5P_3O_{10}$ | 35 | 45 | 40 |
| $Na_2SiO_3 \cdot 5H_2O$ | 11 | 10 | 11 |
| $Na_2CO_3$ | 25 | 23 | 20 |
| $Na_2SO_4$ | 20 | 10 | 14 |

No. 7

| | |
|---|---|
| "Plurafac" B-26 or D-25 | 10.0 |
| "Carbose" D (CMC) | 0.6 |
| KOH (50% solution) | 4.0 |
| $K_4P_2O_7$ (60% solution) | 36.0 |
| "Gantrez" AN-149 | 0.6 |
| $Na_2SiO_3$ | 5.8 |
| Water | 43.0 |

For best results, combine the ingredients as follows: mix the water, 0.01% of the "Plurafac", and the "Gantrez" AN-149 (in that order) with gentle agitation. Heat and maintain this mixture at 80°C. (176°F) for 20-30 minutes or until it becomes more fluid. With gentle agitation, add the potassium hydroxide solution. When the mixture has become homogenous, add the "Carbose"; when it has dissolved, add the sodium silicate. Cool the solution to 60°C (140°F) and with vigorous agitation, add the balance of the surfactant. Continuing the vigorous agitation, rapidly add the tetrapotassium pyrophosphate solution (preheated to 60°C). Optical brighteners, perfumes and dyes may be added immediately and the vigorous agitation continued for another five minutes. The final product can then be cooled and withdrawn into suitable containers.

No. 8

| | |
|---|---|
| "Onyxol" 336 | 5.0 |
| "Ultrawet" 35 KX | 40.0 |
| Sodium xylene sulfonate | 4.0 |
| "Maprofix" SXS | 4.0 |
| Tetrapotassium pyrophosphate | 20.0 |
| Water | 23.5 |

Mix "Ultrawet", SXS and 336. Dissolve phosphate in water at room temperature, add to mixture with agitation. Add metasilicate, 2 oz/gal.

No. 9

| | |
|---|---|
| Water | 57.91-63.41 |

| | |
|---|---|
| "Igepal" CO-730 (1% aqueous sol.) | 1.00 |
| "Gantrez" AN-149 | 0.99 |
| Potassium hydroxide (50% sol.) | 1.80 |
| "Blancophor" RG-96 | 0.10 |
| CMC (low viscosity) | 0.50 |
| Sodium silicate (100%) | 3.70 |
| Colorant | trace |
| Perfume | trace |
| Benzotriazole (tarnish inhibitor) | 0.50 |
| "Tergitol" 15-S-7 (1) | 10.00-13.50 |
| NTA $Na_3$ Crystals | 18.00-12.00 |

Heat water to 80°C and add CO-730, "Gantrez" and potassium hydroxide in order. Agitate with moderate speed and high shear. Use no baffles. Add remaining ingredients in order, maintaining temperature at 60-6P°C, with agitation. (With the exception of "Gantrez" the other ingredients may be added as solutions, with adjustment for water content as necessary.)

This heavy-duty liquid detergent has a 36.6 to 42.0% activity and a pH near 12. It offers the advantages of complete biodegradability and no phosphates for stream enrichment. The emulsion has excellent stability, good freeze-thaw stability, high cleaning efficiency, and freedom from odor.

No. 10

(Intermediate Foamer)

| | |
|---|---|
| "Ucane" 12 or 13 (sulfonate, sodium salt) | 12.0 |
| "Tergitol" 15-S-9 | 4.0 |
| $NTANa_3$ Crystals | 10.3 |
| Sodium tripolyphosphate | 40.0 |
| CMC | 0.5 |
| Sodium silicate | 12.0 |
| Sodium sulfate | 11.0 |
| Optical brightener | 0.3 |
| Soda ash | 3.5 |
| Perfume, fluorescent, and inhibitor (3) | trace to 1.5 |
| Water | 6.4 |

This $NTANa_3$-containing formulation will give whiter washes and lift out soil more effectively than conventional products which contain only conventional builders. Prepare by combining inorganic salts and $NTANa_3$. Add "Tergitol" and sulfonated "Ucane" and blend thoroughly. Then blend CMC and other ingredients into mixture.

### Heavy Alkali Cleaner

For highly alkaline cleaners such as white-wall tire cleaners, liquid steam cleaners, garage floor cleaners, etc., the following concentrate can be marketed:

| | |
|---|---|
| "Varion" 1017 | 10% |
| Sodium metasilicate pentahydrate ("Metso" 20) | 20% |
| Sodium hydroxide | 5% |
| E.D.T.A. sequestrant (40%) | 5% |
| Water | 60% |

### Internal Cleaner for Rolls Used in Plastic Extrusion and Paper Making

| | |
|---|---|
| "Hamp-Ene" 100 | 46.9 |
| "Hamp-Onate" 35 | 6.5 |
| "Disperse" A10-25% | 0.6 |
| Dilute sulfuric or citric acid | to pH 8.0 or 9.0 |
| Water | 46.0 |

Prepare by adding the active ingredients to water with mild agitation. Adjust product with acid to pH 8.0-9.0. Apply by diluting 50 gal of the concentrate in 75 gal of water. Stir and bring solution to a temperature of approximately 140°F. Recirculate mixture through the steam or water connections that lead to the internal sections of the rolls to be cleaned. Continue until rolls are cleaned or solution color remains dark and constant for a few hours. Dump solution and flush rolls with water for up to 30 min. Repeat as required.

### Mop-Treated Compound, Oil-Based

| | |
|---|---|
| Mineral seal oil | 93 |
| "Triton" X-207 | 5 |
| "Hyamine" 3500W80% | 1 |
| Paraffin wax | 1 |

Use full strength. Do not dilute. Label may suggest reduction of bacteria on treated mop.

### Soak Tank Cleaner

| | |
|---|---|
| "Triton" QS-44 | 1.25 |
| "Triton" X-100 | 1.00 |
| Sodium hydroxide | 40.00 |
| Sodium metasilicate anhydrous | 31.50 |
| Soda ash | 26.25 |

### Steam Cleaner

| | |
|---|---|
| "Naxonate" 41 | 2-3 |
| Sodium metasilicate | 15-20 |

| | |
|---|---|
| "Gafac" RE-610 | 5 |
| Water | 73-79 |

---

**Abrasive Cleaner, Liquid**

| | |
|---|---|
| Fine silica | 50.00 |
| "Triton" X-102 | 2.50 |
| Sodium tripolyphosphate | 2.50 |
| "Acrysol" ASE-108 | 6.37 |
| Sodium hydroxide (10% solution) | 4.30 |
| Water, to make | 100.00 |

---

## METAL CLEANERS

**Aluminum Cleaner Brightener**

| | |
|---|---|
| Phosphoric acid (85%) | 47.2 |
| "Triton" X-100 | 2.0 |
| Butyl "Cellosolve" | 16.0 |
| Water | 34.8 |

---

**Aluminum and Stainless Steel Cleaner, Acid**

| | |
|---|---|
| Phosphoric acid (85%) | 3.0 |
| Citric acid, anhydrous | 4.0 |
| "Triton" X-100 | 2.0 |
| Methyl ethyl ketone | 3.0 |
| Water | 88.0 |

Add acids slowly to the water; then add methyl ketone and "Triton" X-100. If cold water is used, the "Triton" X-100 should be premixed by adding to 3 parts warm water before adding.

**Aerosol Chrome Cleaner and Polish**

| | |
|---|---|
| "Union Carbide silicone" L-45, 350 cstks | 2.0 |
| Pumice | 15.0 |
| "Snow Floss" powder | 10.0 |
| Morpholine | 1.5 |
| Water | 42.5 |
| Oleic acid | 2.0 |
| Pine oil | 0.5 |
| Naphthol mineral spirits | 26.5 |

To 100 parts of the above formula add:

| | parts |
|---|---|
| Triethanolamine lauryl sulfate "Hyonic" | 2.5 |
| "Hyonic" FA-40 | 0.5 |
| "Carbopol" 934 (B. F. Goodrich), 2% aqu. sol. | 5.0 |

Dissolve "Union Carbide" L-45, oleic acid, and pine oil in naphthol mineral spirits with simple agitation. Dissolve morpholine in water. Add with pumice and "Snow Floss" powder slowly to the water-morpholine solution with rapid agitation to make a uniform slurry. Then add the naphthol mineral spirits solution slowly while continuing rapid agitation. To 100 parts of this formulation add triethanolamine lauryl sulfate and "Hyonic" FA-40 and blend thoroughly. Add the 2% "Carbopol" solution and again blend thoroughly.

Concentrate should be packaged in seamless lacquered tin plate cans with standard foam valve, using "Ucon" 12 propellant, 18.5 percent by weight based on total polish weight.

---

**High-Pressure Car Wash**

| | % By weight | | |
|---|---|---|---|
| | No. 1 | No. 2 | No. 3 |
| $Na_5P_3O_{10}$ | 80-83 | 67-70 | 55-58 |
| Filler (Puffed borax or sodium sulfate) | — | 10 | 20 |
| $Na_2SiO_3$ (Spray-dried) | 2-5 | 2-5 | 2-5 |
| "Plurafac" D-25 | 15 | 18 | 20 |

The recommended use concentration for high-pressure spray applications is from 0.3 to 0.8%. If a liquid concentrate is desired, a 3-8% solution can be prepared readily for further proportioned dilution at 10:1. The basic formulations can be adapted to water hardness of individual areas by variation of the sodium tripolyphosphate and the filler.

---

**Car Wash Detergent**

| | |
|---|---|
| STPP | 85 |
| "Triton" X-114 | 15 |

The free-flowing powder is premixed at a rate of 0.5 lb/gal of $H_2O$. The gallon of detergent mixture is added to 25 gal of water for the actual washing procedure.

---

**Car Wash-Wax**

| | |
|---|---|
| "Rokyal" 60-T (an LAS amine salt) | 20 |
| Sodium xylene sulfonate (anhydrous) | 3 |
| "Triton" X-100 | 2 |
| "Rowax" 100 | 3 |
| "Versene" 100 | 1 |
| Methanol or isopropanol | 2 |
| "Loterge" AB-30 | 2 |
| Water | 67 |

Mix the "Rokyal" 60-T with sodium xylene sulfonate and one half of the water. Heat mixture to 65-70 C. Agitate to a state of uniformity. Add the remaining water with agitation. Add the remaining products in the order listed, stirring until uniformity is obtained.

---

**Metal Cleaner, Light Duty**

No. 1

| | |
|---|---|
| "Triton" X-100 | 5.00 |
| "Pilot" ABS-99 | 4.00 |

| | |
|---|---|
| Sodium metasilicate | 3.00 |
| TKPP | 3.00 |
| Butyl "Cellosolve" | 5.00 |
| "Tamol" SN | 1.00 |
| Water | 79.00 |

No. 2

| | |
|---|---|
| Sodium metasilicate (pentahydrate) | 34.50 |
| Sodium phosphate (monobasic) | 12.00 |
| Trisodium phosphate (dodecahydrate) | 33.50 |
| "Triton" X-100 | 5.20 |
| "Triton" QS-15 | 2.95 |
| Soda ash | 11.85 |

No. 3 (Acid)

| | |
|---|---|
| Phosphoric acid (85%) | 35.00 |
| Hydroxyacetic acid | 1.00 |
| "Triton" X-100 | 1.50 |
| Water | 62.50 |

Add phosphoric acid and hydroxyacetic acid to water. Add "Triton" X-100 next. If cold water is used, preblend "Triton" X-100 by adding to 3 parts warm water before adding.

No. 4 (Heavy-Duty)

| | |
|---|---|
| Water | 80.10 |
| Sodium hydroxide | 0.60 |
| "Triton" X-102 | 2.00 |
| Alkyl aryl sulfonic acid (98% active) | 2.90 |
| Tripotassium phosphate | 2.20 |
| Sodium nitrite | 0.20 |
| "Tamol" SN | 2.00 |
| Butyl "Cellosolve" | 10.00 |

Add all ingredients except butyl "Cellosolve". When all ingredients are thoroughly dispersed, then add butyl "Cellosolve".

No. 5 (Heavy-Duty, Controlled Foam)

| | |
|---|---|
| "Triton" CF-32 | 5.00 |
| Sodium hydroxide | 32.00 |
| Sodium metasilicate (anhydrous) | 32.00 |
| Soda ash | 31.00 |

**Tank Cleaner**
(Soak Type)
No. 1

| | |
|---|---|
| "Plurafac" RA-20 | 2 |
| $Na_2CO_3$ | 47 |
| $Na_3PO_4$ | 15 |
| $Na_5P_3O_{10}$ | 20 |
| Borax | 15 |
| LAS | 1 |

An effective alkaline soak-type cleaner (used in concentrations of 4-8 oz/gal of water) is:

No. 2

| | |
|---|---|
| "Plurafac" B-26 | 3 |
| $Na_2SiO_3$ | 40 |
| NaOH | 10 |
| $Na_4P_2O_7$ | 7 |
| $Na_3PO_4$ | 40 |

No. 3

| | |
|---|---|
| "Triton" QS-44 | 1.25 |
| "Triton" X-100 | 1.00 |
| Sodium hydroxide | 40.00 |
| Sodium metasilicate, anhydrous | 31.50 |
| Soda ash | 26.25 |

---

**Steam Cleaning Compound**

| | |
|---|---|
| Sodium hydroxide | 30 |
| Borax | 30 |
| Soda ash | 20 |
| Disodium phosphate | 15 |
| "Triton" QS-15 | 5 |

Heat water to 140°F before adding.

---

**Automotive Parts Cleaners**

These cleaners are used for removing grease, carbon and resin-like coatings from automotive engine parts such as carburetors, valves and other functional rods, gears, etc. It is current practice to lower the parts to be cleaned into the cleaning solution in an oversized 5-gal pail, using a perforated basket. Loss of the volatile ingredients is kept to a minimum by a water blanket which floats on the heavier solvent layer. Formulations therefore must contain sufficient chlorinated or other heavier than water solvents to ensure that the specific gravity of the solvent mix is appreciably heavier than water. An effort is being made to replace the high-odor cresylic acid with other low-odor solvents. The following formula can be taken as typical of this type of product.

| | |
|---|---|
| Methylene chloride | 41.0 |
| Orthodichlorobenzene | 20.0 |
| Cresylic acid | 20.0 |
| Tall oil | 0.8 |
| 25% Potassium hydroxide solution | 0.4 |
| Water for tall oil and alkali mix | 5.0 |
| Sodium chromate (Antirust) | 0.2 |
| Water | 12.6 |

Instead of using a tall-oil/alkali premix, the formulator can add the potassium hydroxide solution to the water and the tall oil to the solvent mix to give across-the-boundary neutralization.

---

**Tanker Cleaner (Concentrate)**

| | |
|---|---|
| Kerosene | 74.0 |
| "Triton" X-45 | 4.5 |
| "Triton" X-100 | 12.0 |
| Butyl "Cellosolve" | 9.5 |

Add the Triton X-100 and Triton X-45 to the butyl "Cellosolve" and then add to the kerosene.

---

**Emulsion Degreaser for Metal**

| | |
|---|---|
| Kerosene | 3 |
| "Emcol" P10-59 | 1 |

This concentrate dilutes 1 to 4 or 1 to 9 with kerosene at time of use.

---

**Radiator Cleaner**

A combination of 3-5 parts by weight of "Plurafac" A-24 with 95-97 parts by weight of oxalic acid will effectively clean automotive radiators. It removes rust, suspends scale and prevents the cooling system from clogging. Unlike anionic surfactants, "Plurafac" A-24 is a low foamer.

---

**Solvent Emulsion Engine Block Cleaner**

| | |
|---|---|
| "Triton" GR-7 | 5 |
| Kerosene | 45 |
| Heavy aromatic naphtha | 40 |
| Cresylic acid | 10 |

Paint on engine block. Allow to stand for a time, then flush surface with water or steam.

---

**Solvent Emulsion, Sludge and Carbon Cleaner**

| | |
|---|---|
| Kerosene | 57.5 |
| Cresylic acid | 20.0 |
| Heavy aromatic naphtha | 10.0 |
| Orthodichlorobenzene | 10.0 |
| "Triton" X-102 | 2.5 |

---

**Machinery Steam Cleaner, Liquid**

| | |
|---|---|
| "Triton" H-55 | 2.0 |
| "Triton" X-114 | 0.4 |
| KOH | 14.1 |
| Sodium silicate (1:1.8) | 15.2 |
| TKPP | 17.2 |
| Water | 51.1 |

Use 0.3 to 0.5% in water.

---

**Cold Tank Cleaner and Degreaser**

5-10% of "Surfactol" 365 produces an emulsifiable form of orthodichlorobenzene or other chlorinated chemicals to produce excellent solvents for resins, gums, heavy greases, tars and carbonaceous materials.

---

**Dairy Pipeline Cleaner**

No. 1

| | |
|---|---|
| "Triton" CF-10 | 3 |
| Caustic soda | 20 |
| Sodium metasilicate | 30 |

| | |
|---|---|
| Sodium tripolyphosphate | 47 |

No. 2
(Low Foam)

| | |
|---|---|
| "Triton" CF-54 | 5 |
| NaOH | 10 |
| Sodium silicate (anhydrous) | 30 |
| Soda ash | 30 |
| Sodium tripolyphosphate | 25 |

---

**Boiler Descalant**

An effective descalant for boilers can be formulated with 2% of "Plurafac" A-24 based on the weight of sulfamic acid used. If necessary, an inhibitor may be added at a concentration of 0.5% of acid weight.

---

**Degreaser, Pots and Pans**

| | |
|---|---|
| "Triton" X-100 | 20 |
| "Triton" X-301 | 50 |
| Water | 30 |

If cold water is used, preblend the "Triton" X-100 with 3 parts warm water before adding.

---

**Dental-Instrument Cleaner-Sanitizer**

| | |
|---|---|
| "Hyamine" 3500—50% | 23 |
| "Triton" X-165—70% | 27 |
| Sodium nitrite | 15 |
| Water | 35 |

**Detergent Sanitizer**

No. 1

| | |
|---|---|
| "Hyamine" 3500 | 10 |
| "Triton" X-100 | 5 |
| *Methyl cellulose (to thicken) | 1 |
| Water | 84 |

No. 2

| | |
|---|---|
| "Hyamine" 2389 (50%) | 10 |
| "Triton" X-100 | 10 |
| *Methyl cellulose (to thicken) | 1 |
| Water | 79 |

* "Methocel" 1500 cps or "Cellosize" QP-4400.

No. 3

| | |
|---|---|
| "Hyamine" 2389 (50%) | 5.5 |
| "Triton" X-100 | 10.0 |
| Pine oil | 20.0 |
| Water | 64.5 |

No. 4

| | |
|---|---|
| "Hyamine" 2389 (50%) | 5.5 |
| "Triton" X-100 | 10.0 |
| "Solvenol" No. 2 | 5.0 |
| Pine oil | 15.0 |
| Water | 64.5 |

No. 5

| | |
|---|---|
| "Hyamine" 2389 (50%) | 10.00 |
| "Triton" X-100 | 10.00 |
| Pine oil | 2.50 |
| Isopropanol | 0-10 |
| Trisodium phosphate | 0.25 |
| Dye | 0.05 |
| Water | 67.20 |

No. 6

Acid Detergent-Sanitizer

| | |
|---|---|
| "Hyamine" 3500 (50%) active | 20 |
| "Triton" X-114 | 10 |
| Phosphoric acid | 20-34 |
| Water | balance |

Cleaner-sanitizer dilution, ¼ oz per gal; disinfectant dilution, ½ oz per gal.

No. 7

| | |
|---|---|
| "Plurafac" B-26 or D-25 | 10 |
| $Na_5P_3O_{10}$ | 30 |
| $Na_2SiO_3$ | 20 |
| $Na_2CO_3$ | 35 |
| Quarternary germicide | 5 |

No. 8

| | |
|---|---|
| "BTC"-2125 (50%) | 12.8 |
| "Neutronyx" 656 | 8.0 |
| Sodium carbonate | 4.0 |
| STPP | 2.0 |
| "Seqonyx" 100 | 1.0 |
| Water, to make | 100.0 |

Charge water into clean vessel and warm to 40-45°C; slowly add soda ash until completely dissolved. Add STPP in same manner. Add "BTC"-2125, "Seqonyx" and "Neutronyx" in any order. Package.

Use 1 oz/per gal of water for disinfection.

No. 9

| | |
|---|---|
| "BTC"-2125 | 25 |
| "Ammonyx" MCO | 17 |
| STPP | 5 |
| Soda ash | 5 |
| Water, to make | 100 |

No. 10 (Acid)

| | |
|---|---|
| "Hyamine" 3500—50% | 10 |
| "Triton" X-114 | 10 |
| Phosphoric acid | 20 |
| Water | 50 |

Use 1 oz per 4 gal of water.

No. 11

| | |
|---|---|
| "Variquat" 415 | 10 |
| Nonionic detergent | 8 |
| Tetrapotassium pyrophosphate | 5 |
| Water | 77 |

No. 12

| | |
|---|---|
| "Triton" X-100 | 4.0 |
| "Hyamine" 3500 | 1.5 |
| TKPP | 7.5 |
| Water, to make | 100.0 |

For sanitizing and cleaning, use at rate of 2 oz per gal of water. For disinfecting use at rate of 3½ oz per gal of water.

---

**Detergent-Sanitizer Deodorant**

| | |
|---|---|
| "Hyamine" 2389 | 10 |
| "Triton" X-100 | 10 |
| STPP | 2 |
| "Versene" | 1 |
| Water | 77 |

Perfume and dye as desired

Use 1 oz per 2 gal of water.

**Sanitizers**

| | No.1 | No. 2 |
|---|---|---|
| "Hyamine" 3500 (active) | 10 | 10.0 |
| "Triton" X-100 | — | 2.5 |
| EDTA | — | 0.1-0.25 |
| Water, to make | 100 | 100 |

Use 1 oz to 4 gal water.

No. 3 (Restaurant)

Used in rinse water to drastically reduce bacterial counts on utensils.

| | |
|---|---|
| "Variquat" 415 | 10 |
| "Versene" 100 | 2 |
| Water | 88 |

Use 1 oz per 2 gal rinse water. Follow with clear water rinse.

No. 4 (Wash Room)

| | |
|---|---|
| "Barquat" M B 50 | 5.00 |
| EDTA sodium salt | 0.75 |
| "Surfonic" N 95 | 5.00 |
| "Surfonic" N 40 | 5.00 |
| Butyl "Carbitol" | 2.00 |
| Tetrapotassium pyrophosphate | 3.00 |
| Water | 79.25 |

Usually a color and perfume must be included in the formulation for consumer appeal. Choice of a color poses no problem, but the perfume must be selected for odor value and compatibility with other ingredients. Where a larvicide is to be included, such as an aromatic petroleum derivative, it must be fitted into the formula. Where a clear product is desired, adjustment of the coupling agent (butyl "carbitol') may be required.

**Pine-Oil Disinfectant (Emulsifiable)**

| | |
|---|---|
| "Hyamine" 2389 (50%) | 6.4 |
| "Triton" X-45 | 5.0 |
| "Dytol" B-35 | 5.0 |
| Pine oil | 3.0 |
| "Solvenol" No. 2 | 9.0 |
| "Coprol" B (Naplthenic oil) | 71.6 |

Phenol coefficient: *Salmonella typhosa* — 10, *Staphylococcus aureus*, 12.

Use dilution: Sanitizing, 1 oz per gal of water; disinfecting, 1.75 oz per gal of water.

**Pine-Type Detergent-Sanitizer**

No. 1

| | |
|---|---|
| "Hyamine" 2389 (50%) | 10.00 |
| "Triton" X-100 | 100.00 |
| Pine oil | 2.50 |
| Isopropanol | 10.00 |
| TSP | 0.25 |

| | |
|---|---|
| Dye | 0.05 |
| Water | 67.20 |

Predissolve dye in "Triton" X-100 before adding. Use 1 oz per 2 gal of water.

No. 2

| | |
|---|---|
| Water | 64.5 |
| "Triton" X-100 | 10.0 |
| "Hyamine" 2389 | 5.5 |
| "Solvenol" No. 2 | 5.0 |
| Pine oil | 15.0 |

Phenol coefficient: *Salmonella typhosa*, 5.5; *staphylococcus aureus*, 9.0.

Use dilution: Disinfecting, 2 oz per gal of water; cleaning and sanitizing, 1 oz per gal of water.

---

**Toilet-Bowl Cleaner**

Calculated phenol coefficient of 7-9 against *S. typhosa*. Stable for at least 10 weeks at 0°, 70° and 120°F.

| | |
|---|---|
| "Hyamine" 3500 (50%) | 5 |
| "Triton" X-114 | 2 |
| Phosphoric acid (75%) | 30 |
| Water | 63 |

Add phosphoric acid slowly to the water. Then add "Triton" X-114 and "Hyamine" 3500.

No. 2

Suitable for packaging in plastic bottles.

| | |
|---|---|
| "Triton" X-305 | 3.25 |
| HC1 (active) | 18.00 |
| "Rhoplex" X-52 | 2.50 |
| Water | 76.25 |

Mix "Triton" X-305 with 2/3 of the water. Add this to the acid slowly with mixing. Mix the "Rhoplex" X-52 with the remaining water and add very slowly with mixing.

No. 4

May be packaged in polyethylene.

| | |
|---|---|
| "Triton" X-100 | 0.5 |
| "Hyamine" 2389 (50%) | 5.0 |
| Hydrochloric acid (active) | 14.0 |
| Water | 80.0 |
| Pine oil | 0.5 |

Add the hydrochloric acid to the water slowly; then add remaining ingredsients in any order desired. Use 1 oz of product per bowl.

No. 5

| | |
|---|---|
| "Hyamine" 3500 (50%) | 0.10 |
| "Triton" N-101 | 0.84 |
| "Triton" X-114 | 0.84 |
| "Morton Latex" E-153 | 1.30 |
| Water | 19.40 |
| Muriatic acid 2·° Baume (31.5% active) | 77.52 |

To disinfect, pour 2 oz of bowl cleaner into water remaining in cleaned bowl. Swab entire bowl thoroughly with solution. Flush toilet. For ease of preparation dilute the "Latex" E-153 with part of the water before blending.

No. 6

This acid bowl cleaner should not be packaged in polyethylene as the orthodichlorobenzene will bleed through the container. Glass containers are recommended.

| | |
|---|---|
| "Triton" X-155 | 2.5 |
| Orthodichlorobenzene | 2.5 |
| "Rhoplex" X-52 | 0.5 |
| Muriatic acid (18° Baumé) | 65.0 |
| Water | 29.5 |

Mix "Triton" X-155 and orthodichlorobenzene well; then add 2/3 of the water and mix thoroughly. The muriatic acid is then added slowly and mixed well. The "Rhoplex" X-52 is diluted with the remaining 1/3 of the water. The final step is the addition of the diluted "Rhoplex" solution with stirring until the mix assumes a uniform milky-white appearance.

**Powdered Dairy Sanitizer**

Used for cleaning and sanitizing milking equipment in one operation.

| | |
|---|---|
| "Variquat" 415 | 10 |
| Nonionic detergent | 5 |
| Sodium tripolyphosphate | 25 |
| "Versene" powder | 2 |
| Soda ash | 58 |

Spray the liquids into the powders slowly while tumbling in a ribbon blender. A few percent of water sprayed in at the end helps agglomerate the powder and gives a less tacky product.

**Dairy Liquid Germicide and Milkstone Remover**

Milder acids such as gluconic or hydroxyacetic may be substituted for the phosphoric acid in this formulation. "Hyamine" 2389 (50%) may be substituted for "Hyamine" 3500 (50%).

| | |
|---|---|
| "Hyamine" 3500 (50%) | 16 |
| "Triton" X-114 | 5 |
| Phosphoric acid (75%) | 30 |
| Water | 49 |

Utensils should be soaked in a solution containing from 1/3 to 1 oz of acid detergent-sanitizer in a gallon of water and thoroughly brushed and rinsed.

**Low-Foam Dairy Pipeline Cleaner**

| | |
|---|---|
| "Triton" CF-54 | 5 |
| NaOH | 10 |
| Sodium silicate (anhydrous) | 30 |
| Soda ash | 30 |
| Sodium tripolyphosphate | 25 |

**Egg Cleaner-Sanitizer**

No. 1

| | |
|---|---|
| $Na_2CO_3$ | 21 |
| $Na_4P_2O_7$ | 37 |
| $Na_2SiO_3 \cdot 5H_2O$ | 10 |
| $Na_3PO_4$ (chlorinated) | 22 |
| "Plurafac" RA-43 | 5 |

| | |
|---|---|
| Water | 5 |

1. Spray a mixture of "Plurafac" and the water onto the tetrasodium pyrophosphate (alone or mixed with other inorganic salts) with continuous mixing, whereby hydration of the phosphate and simultaneous absorption of the surfactant occur.
2. Add the hydrated sodium metasilicate to the hydrated phosphate, still mixing.
3. Add the active chlorine-containing compound to the above mixture and continue mixing until homogenized and a dry, free-flowing, granular product is obtained. (The total amount of water used should not exceed the amount shown in the formulations).

Maximum chlorine stability is obtained when the Plurafac surfactant, diluted with water, is absorbed by the phosphate mixture (one of those phosphates must be tetrasodium pyrophosphate). Other organic salts (e.g., sodium tripolyphosphate, sodium carbonate, sodium sulfate) may be included, provided at least 50% of the mixture is tetrasodium pyrophosphate. Step 1 avoids temperature-control problems and eliminates the need for an aging period.

In Step 2, hydrated metasilicates are used because they do not promote additional agglomeration. It is essential that the metasilicate be added after the phosphate ingredient has hydrated and absorbed the surfactant to avoid discoloration and degradation.

Step 3: Add the chlorine-containing compound. Active chlorine-containing compounds which may be used in this process include chlorinated trisodium phosphate, trichlorocyanuric acid, the sodium or potassium salt of dichlorocyanuric acid, dichlorodimethylhydantoin ("Wyandotte Halane") and other organic active chlorine-containing compounds.

No. 2

This product has good stability. Storage for 3 hr at 30 to 32°F does not freeze or thicken the product.

| | |
|---|---|
| "Hyamine" 3500 (50%) | 20 |
| "Triton" X-100 | 10 |
| TSP | 3 |
| Sodium metasilicate | 4 |
| Water | 63 |

Use 1 oz per 4 gal of water. Very dirty eggs may require a stronger solution.

No. 3

| | |
|---|---|
| "Triton" X-100 | 5 |
| "Hyamine" 3500 (50%) | 6 |
| Kerosene | 2 |
| TSP anhydrous | 4 |
| Distilled water | 83 |

Preblend "Triton" X-100 and kerosene before adding to water.

Use 1 oz per 2 gal of water. Very dirty eggs may require a stronger solution.

No. 4

| | |
|---|---|
| "Hyamine" 2389 (50%) | 10 |
| "Triton" CF-32 | 5 |
| Soda ash | 55 |
| Sodium metasilicate | 20 |
| Sodium tripolyphosphate | 10 |

Under conditions of severe agitation, this formulation will not be satisfactory because of excessive foam generation.

No. 5

(Hand-washing or Immersion Machines)

| | |
|---|---|
| Sodium tripolyphosphate | 50 |
| Sodium metasilicate, pentahydrate | 15 |
| Soda ash, light density | 20 |
| "Hyamine" 2389 (50%) | 10 |
| "Triton" X-100 | 5 |

No. 6

(For Spray or High-Pressure Machines)

| | |
|---|---|
| Sodium tripolyphosphate | 47 |
| Sodium metasilicate pentahydrate | 14 |
| Soda ash, light density | 19 |
| "Hyamine" 2389 (50%) | 10 |
| "Triton" CF-21 | 5 |
| Light mineral oil (70 vis) | 5 |

---

**Disinfecting Vegetable Sacks**

Soak bags for one hour in

| | |
|---|---|
| "Hyamine" 2389 | 1 oz |
| Water | 2 gal |

---

**Aerosol Disinfectant Spray**

| | |
|---|---|
| "BTC"-2125 | 0.4 |
| Triethylene glycol | 5.0 |
| Propylene glycol | 5.0 |
| Water | 4.0 |
| Isopropanol | 63.5 |
| Propellant | q.s. |

**Swimming Pool Algaecide**

Used for preventing growth of unsightly scum:

No. 1

| | |
|---|---|
| "Variquat" 415 | 10 |
| Water | 90 |

As in all quaternary formulations, it is advisable to use softened water for making up this formulation. For a 50,000-gal swimming pool, it is recommended that 1 gal of product be added at the start, then another quart every few days.

No. 2

| | |
|---|---|
| "Hyamine" 2389 | 10 |
| Water | 90 |

Use 10 ppm of water.

**Barber Shop Tool Disinfectant**

| | |
|---|---|
| "Hyamine" 2389 (50%) | 4.00 |
| Isopropyl alcohol | 16.00 |
| "Versene" (salt No. 2) | 0.34 |
| Water | 79.66 |

Use 5 oz per 1 gal of water.

**Thermometer Sterilizer**

| | |
|---|---|
| "Hyamine" 1622 (Crystals) | 0.5 |
| Alcohol | 50.0 |
| Acetone | 10.0 |
| Water | 39.5 |

Dissolve the "Hyamine" 1622 Crystals in water before adding alcohol and acetone.

Use full strength.

**White-Wall Tire Cleaner**

No. 1

| | |
|---|---|
| "Plurafac" A-38 | 5 |
| $Na_2SiO_3$ | 5 |
| $Na_2CO_3$ | 5 |
| $Na_2SO_4$ | 85 |

In using, scrub the powder on the wet tirewall with a damp brush. A liquid solution (4-8 oz/gal) can also be brushed or sponged onto the tire. After application, flush the tire thoroughly with water.

No. 2

| | |
|---|---|
| "Emcol" 5130 | 7 |
| Trisodium phosphate | 4 |
| Low-salt alkyl aryl sulfonate | 1 |
| Water | 88 |

No. 3

| | |
|---|---|
| Sodium metasilicate, anhydrous | 5.0 |
| Trisodium phosphate, anhydrous | 5.0 |
| "Triton" QS-44 | 0.9 |
| "Triton" X-100 | 1.3 |
| Water | 87.8 |

No. 4

| | |
|---|---|
| 1. "Igepal" CO 630 | 20 |
| 2. Sodium metasilicate | 10 |
| 3. Sodium xylene sulfonate | 20 |
| 4. Water | 60 |

5. Add 0.2% "Pontamine" B. T. White or similar optical whitener.

Dissolve 2 and 3 in the water. Add 1 and stir until a uniform solution results. Add 5 and stir until dissolved.

**Paint Stripper**

No. 1

| | |
|---|---|
| "Triton" QS-44 | 1.5 |
| Sodium hydroxide | 15.0 |
| Water | 83.5 |

No. 2

| | |
|---|---|
| "Naxonate" 4L | 2-3 |
| Sodium hydroxide | 15 |
| "Gafac" RE-610 (GAF) | 5 |
| Water | 78-79 |

**Wax Stripper**

| | |
|---|---|
| "Triton" QS-44 | 3.2 |
| Sodium hydroxide | 1.7 |
| Tetrapotassium pyrophosphate | 12.0 |
| Tetrasodium pyrophosphate | 7.0 |
| Sodium metasilicate, anhydrous | 0.4 |
| Water | 75.7 |

---

**Solvent-Emulsifier, Oil Slick**

No. 1

| | |
|---|---|
| "Triton" X-45 (or N-57) | 20 |
| "Solvesso" 150 | 80 |

No. 2

| | |
|---|---|
| "Triton" N-101 | 12 |
| Nonylphenol | 8 |
| "Solvesso" 150 | 80 |

The solvent-emulsifier is sprayed onto the oil slick to be removed, allowed to react for a few minutes, then hosed off if removal from metal, wood or concrete is desired. If dispersion in water is necessary, such as oil spills in harbor or dock areas, agitation with a motor boat is recommended.

CHAPTER VII

# EMULSIONS

### Mineral Oil Emulsion

| | |
|---|---|
| "Emcol" 5138 | 2.0 |
| Mineral oil | 18.0 |
| Water | 80.0 |

Add "Emcol" 5138 to half of water. Add mineral oil with stirring. Add balance of water with stirring.

### Paraffin Oil Emulsion

| | |
|---|---|
| "Gantrez" AN-169 | 1 |
| Paraffin oil (No. 1) | 3 |
| Monoethanolamine | 1 |
| Water | 95 |

Mix copolymer and oil until lump-free. Add to water-ethanolamine solution and blend until smooth.

### Kerosene Emulsion

| | |
|---|---|
| "Gantrez" AN-169 | 3 |
| Kerosene | 30 |
| Water | 67 |

Dissolve copolymer in water, add kerosene and mix.

### Soluble Kerosene

| | |
|---|---|
| "Emcol" P10-59 | 10-14 |
| Kerosene | 28-32 |
| Water | 62-54 |

### Silicone Water-Repellant Emulsion

| | |
|---|---|
| "Emcol" 5130 | 3 |
| Water | 68 |
| Dow Corning "Decetex" 104 | 29 |

Dissolve "Emcol" 5130 in about half the water. Add "Decetex" 104 with vigorous mixing until thick phase is homogenous. Add balance of water and mix well. Resulting 25% silicone emulsion is diluted approximately 1 to 10 with water before applying.

### "PICCOLLOID" Emulsion

| | |
|---|---|
| "PICCOLLOID" | 64.40 |
| "Conco" AAS-50-T | 0.33 |
| Water (1) | 6.00 |
| Water (2) | 6.00 |
| Water (3) | 11.04 |
| 2% solution "Cellosize" QP-4400 | 12.23 |

Charge all the "PICCOL-

LOID" solution and "Conco" AAS-50-T to the high-speed mixer and agitate for about 5 min at high speed. Add water (1) a little at a time, allowing each addition to mix in before adding the next and raising the mixing blade as required to maintain an open vortex. Continue this procedure with water (2). Inversion should occur near the end of portion (1) or while adding the first part of (2). After inversion, raise the mixing blade until it is near the surface and add small amounts of water around the blade until the batch begins to roll over again, then add water around the sides of the tank until roll-over becomes rapid. All the remaining water and thickener solution may be added as quickly as desired. Agitate for a few minutes to obtain a homogeneous mixture.

## Hydrocarbon Resin Emulsion

No. 1

| | |
|---|---|
| Resin solution in mineral spirits | 100.0 |
| Soya fatty acids | 10.0 |
| 2-Amino—2-methyl—1-propanol | 3.0 |
| Water | 50.0-100.0 |

No. 2

| | |
|---|---|
| Resin solution in mineral spirits | 100.0 |
| Soya fatty acids | 10.0 |
| Morpholine | 3.0 |
| Water | 50.0-100.0 |

No. 3

| | |
|---|---|
| Resin solution in mineral spirits | 100.0 |
| Rosin | 6.0 |
| Potassium hydroxide | 1.2 |
| Water | 50.0-100.0 |

No. 4

| | |
|---|---|
| Resin solution in mineral spirits | 100.0 |
| Oleic acid or tall oil acids | 6.0 |
| Potassium hydroxide | 1.4 |
| Water | 50.0-100.0 |

Add the oil-soluble ingredients and solvents to the mixer. With high speed agitation, add small increments of solid resin to the solution and completely dissolve. Suitable resin solution concentration and viscosity will depend on the power limitations of the mixer used. Add the amine or KOH to approximately 25% of the total water to be used and add this solution slowly until the change of phase from water-in-oil to oil-in-water occurs. Add the remainder of the water.

---

**Cationic "Castorwax" Emulsion**

| | |
|---|---|
| "Castorwax" | 21.0 |
| "Triton" X-100 | 1.1 |
| "Aerosol" C-61 | 2.6 |
| Water | 75.3 |

Add "Triton" X-100 to one third of the water and heat to 212°F. Add "Aerosol" C-61 to "Castorwax" and heat to 212°F. When the "Castorwax"/"Aerosol" C-61 mix begins to foam or boil, remove from the heat and add slowly to the water with very rapid agitation. The temperature of the water must be maintained above 200°F during the addition of the "Castorwax" so that the temperature, after all the "Castorwax" has been added, remains above 200°F. Continue stirring, allowing the temperature to fall slowly to approximately 170°F. When the temperature has fallen to 170°F the remaining two-thirds of the water can be worked in slowly, with moderate agitation. Emulsions of "Castorwax" containing cationic emulsifiers tend to thicken as they cool after all the water has been added. Homogenize or violently agitate the batch after it is thoroughly cooled, to produce the final low viscosity emulsion.

---

**"Castorwax" Emulsion**

| | |
|---|---|
| "Castorwax" | 10.70 |
| Stearic acid | 2.14 |
| Triethanolamine | 1.07 |
| "Triton" X-100 | 0.53 |
| Water | 85.56 |

Melt the "Castorwax" and stearic acid together and bring to a temperature of about 212°F. At the same time add the triethanolamine and "Triton" X-100 to about one third of the total water and bring to its boiling point. Add the melted "Castorwax"-acid mix slow to the TEA water solution with rapid agitation. While the temperature is still above 200°F, run the emulsion through a colloid mill several times to ensure a small particle size. The remaining water at about 200°F can then be added to the emulsion in the colloid mill. Cool the mixture slowly to room temperature with continued agitation.

---

**Wax-to-Water Cationic Emulsion**

| | |
|---|---|
| "Epolene" E-10 | 40 |
| Cationic surfactant | 8 to 10 |
| Acetic acid | 2 to 4 |

| | |
|---|---|
| Water | 150 |

Melt the "Epolene" E-10 and cationic surfactant and cool to 120°C (250°F). The melt temperature should not exceed 150°C (302°F). Add the acetic acid to the "Epolene" E-10 and cationic surfactant melt and mix thoroughly (mixing time, 1 to 2 min). This step should be closely controlled. Acetic acid boils at about 118°C, and prolonged heating will evaporate the acetic acid and cause faulty emulsification. While maintaining good *but not violent* agitation, slowly pour the melt into water which has been heated to 95-97°C.

Violent or too rapid agitation results in the formation in the liquid of a vortex which quite often exposes the agitator or stirrer. Molten material contacting the exposed agitator may solidify on the surface and then separate from the agitator and be dispersed in the emulsion as solid particles. Dispersion of the melt can be aided by injection of the melt below the surface of the water. This method also minimizes solidification of the melt on the surface of the water. The rate of addition of the melt depends on the size of the melt tank, the size of the water vessel, and the rate of agitation; however, the melt should be added at the same rate at which it is being dispersed in the water. The formation of puddles of molten material on the surface should be avoided.

After emulsification is complete, continue stirring the emulsion until it has cooled to at least 45°C. Rapid cooling of the emulsion will help set the emulsion particle size and produce clearer emulsions.

**Wax to Water Anionic Emulsion**

| | |
|---|---|
| "Epolene" E-10 | 40 |
| Tall oil fatty acids | 7 |
| Morpholine | 7 |
| Water | 150 |

Melt the "Epolene" E-10 and fatty acids; then cool to 120°C (250°F). Add the morpholine to the "Epolene" E-10 melt with stirring, and heat at 120°C (250°F) for one to two minutes.

This step is probably the most critical one in the emulsification of "Epolene" E-10 and should be closely controlled. Morpholine boils at about 128°C (262°F); prolonged heating will evaporate the amine and cause poor emulsification. At this stage, the progress of the reaction can be followed by noting the appearance of the melt. When the amine is first added, the melt is hazy. After the reaction, the melt becomes clear. While maintaining good *but not violent* agitation, slowly pour the melt into water which has been heated to 95-97°C (203-207°F).

Violent or too rapid agitation results in the formation in the liquid of a vortex which often exposes the agitator or stirrer. Molten material contacting the exposed agitator may solidify on the surface and then separate from the agitator and be dispersed in the emulsion as solid particles. Dispersion of the melt can be aided by injection of the melt below the surface of the water. This method also minimizes solidification of the melt on the surface of the water. The rate of melt addition depends on the size of the melt tank, the size of the water vessel, and the rate of agitation; however, the melt should be added at the same rate at which it is being dispersed in the water. The formation of puddles of molten material on the surface should be avoided.

After emulsification is complete, continue stirring the emulsion until it has cooled to at least 45°C (113°F). Rapid coolIng of the emulsion will help set the emulsion particle size and produce clearer emulsions.

---

**Pressure Emulsification**

Using a "Dura-Lydo" pressure emulsification apparatus and DOWA-0355 resin, to make a total of 3,500 g of emulsion at 25% N.V., charge into main kettle, (with Sampling Valve shut).

| | |
|---|---|
| "DOWA" 0355 | 730 |
| Oleic acid | 145 |
| Morpholine | 50 |
| Water at room temperature | 350 |

Replace cover, lower the agitator motor into position for correct alignment with agitator clutch, connect thermostat (if thermostat is not used, short the thermostat connecting wire). Place pressure-holding cylinder on the main tank (via two quick-disconnect valves). Shut off both valves on the main tank cover. Unscrew top of morpholine holding cylinder. Charge holding cylinder with morpholine 95 and replace top for tight seal.

Check tight seal of main kettle cover. All three locking studs must be tight. Start hot-plate. Control light is on while heat is on. Insert thermometer or heat recorder probe into main kettle well. Assure good contact by filling bottom of well with stable heat-transfer liquid (silicone oil is suitable).

Raise main kettle temperature to 300°F. Start agitator (caution: never reverse the agitator direction without first shutting off the motor). Release the morpholine in the holding cylinder into the main kettle by (1) opening the right-side valve to equalize head pressure in the cylinder with that in the main kettle: (2) opening the left-side valve to release the morpholine into the main kettle; and (3) after allowing 2 min transfer time, shut off both valves. Charge water (2130) into water kettle (the closer to boiling, the better). Replace cover of water kettle. Check tight lock on the three locking studs. Insert thermometer or heat recorder probe in the well. Assure good contact by heat-transfer liquid as above. Start heat.

Raise water kettle temperature to 300°F, connect the water kettle with the main kettle by means of the transfer pipe. The following procedure is recommended for making the connection: (1) with the viewing glass of the transfer pipe on the side of the water kettle and all valves tightly shut, connect the pipe's other end with the right-side, quick-disconnect (the higher one of the two, branching off from the pressure gauge). *This must be done first*! (2) now insert the other end of the pipe into the self-closing quick disconnect of the water kettle.

Connect the compressor hose to the water kettle. Start compressor and raise tank pressure to about 125 psi. Continue heat and agitation in the main kettle 30 min after the temperature in the main kettle has reached 300°F. Open compressed-air inlet valve of the water kettle slowly and raise head pressure 15 psi above that of the main kettle. Open water inlet valve of the main kettle. Adjust pressure gradient in the water kettle by repeated feeding of compressed air. When water from the water kettle has been completely transferred to the main kettle, this will be shown by equalization of the pressure in both kettles.

Shut off heat in the water kettle. Continue heat and agitation. Shut off water inlet valve of main kettle. Shut off compressor. Continue heat and agitation for at least 45 min after all water has been transferred. *This is very important*!

Shut off heat in hot-plate under the main kettle. Turn on cooling water. Continue agitation. Cool to 150-160°F. Shut off agitation. Sample. Discharge. Store in covered container.

**Caution:** *Operation of the pressure emulsifier should be entrusted only to trained and skilled personnel.*

CHAPTER VIII

# FARM and GARDEN FORMULATIONS

### Micronutrient Concentrate

| | |
|---|---|
| "Hamp-OL" 9% iron chelate | 55.000 |
| "Hamp-OL" 9% manganese chelate | 27.500 |
| "Hamp-OL" 9% copper chelate | 14.000 |
| "Hamp-ENE" zinc chelate | 9.000 |
| "Solubor" | 2.500 |
| Sodium molybdate | 0.025 |

Dry-blend ingredients and package in containers that prevent moisture pickup and can be resealed if contents will not be rapidly used up. The micronutrient mix can be applied either bimonthly by diluting 1/3 oz (1 level tablespoon) in a gallon of water or mixed with home and garden fertilizers at a rate of 50 to 200 lb per ton.

---

### Antifreeze for Plants

| | |
|---|---|
| PVP/VA S-630 | 1-5 |
| Water, to make | 100 |

Mix well. Spray on plants to prevent excessive transpiration and freezing. Will also prevent wiltage of ornamental plants during transplanting.

---

### Insecticides

| | |
|---|---|
| Lindane | 20.0 |
| "Triton" X-45 | 3.0 |
| "Triton" X-155 | 4.5 |
| Aromatic petroleum solvent | 32.5 |
| Isophorone | 40.0 |

---

### Dormant Spray Oil

| | |
|---|---|
| Oil | 98-99 |
| "Triton" X-45 | 1-2 |

---

### Summer Spray Oil

| | |
|---|---|
| Oil | 95-97 |
| Butyl alcohol | 1.5-3.0 |
| "Triton" X-45 | 1.5-2.0 |

---

### 25% DDT Emulsifiable Concentrate

| | |
|---|---|
| DDT | 25.0 |
| "Triton" X-155 | 3.0 |

| | |
|---|---|
| Xylene | 72.0 |

**1.5 lb/gal Dieldrin Emulsifiable Concentrate**

| | |
|---|---|
| Technical dieldrin | 19.1 |
| "Triton" X-155 | 4.0 |
| "Triton" B-1956 | 2.0 |
| Xylene | 74.9 |

**Pantachlorophenol Emulsion**
(Herbicide)

| | |
|---|---|
| Pentachlorophenol | 25 |
| "Triton" GR-7 | 15 |
| Diacetone alcohol | 25 |
| Heavy aromatic naphtha | 35 |

**Herbicide**

| | |
|---|---|
| Monosodium acid methane arsonate | 4.00 lb |
| "Atplus" 401 | 1.75 lb |
| Water, to make | 1.00 gal |

**Pet Shampoo (Insecticidal)**

No. 1

| | |
|---|---|
| "Miranol" C2M conc. | 25.0 |
| Lauric diethanolamide | 4.0 |
| Hexyleneglycol | 2.0 |
| "Pyrocide" 5192 | 0.5 |
| Water | 68.5 |

No. 2

| | | |
|---|---|---|
| A. | "Veegum" | 1.0 |
| | Water | 42.9 |
| | Citric acid | 0.4 |
| | "Igepon" AC-78 (83% solids) | 18.0 |
| | "Igepon" TC-42 (24% solids) | 25.0 |
| B. | Cetyl alcohol | 1.8 |
| | Glycerol monostearate, acid stable | 5.9 |
| | "Solulan" 98 | 3.5 |
| C. | Synergized pyrethrins (5% piperonyl butoxide & 10% pyrethrins) | 0.5 |
| D. | "Vancide" 89RE | 1.0 |

1. Add the "Veegum" to the water slowly, agitating continually until smooth. Add the rest of A and heat to 75°C.
2. Mix B and heat to 80°C.
3. Add B to A, mixing until cool.
4. Add C to 3 and mix.
5. Add D to a small portion of 4, disperse thoroughly. Add this

concentrate to the remainder of 4. Mix until uniform. The final pH should be about 5.0.

---

**Deer Deterrent**

A farmer has found that spreading manure of bears, tigers and lions about his orchard keeps deer out. He gets this manure from zoos.

---

**Stump Killers**

Effective tree stump killers not only kill the tree but cause its wood and root system to deteriorate. In the old days, holes were drilled into the stump and were filled with potassium nitrate (saltpeter) crystals. In the course of time, the salt permeated the woody structure and rendered it automatically combustible. When the stump had dried out, it was ignited and allowed to burn throughout its root system. The treatment described may be effective for one type of wood (pine) but not for another (eucalyptus). In the writer's experience a large eucalyptus stump is difficult to kill. A stump of this type, 2 ft in diameter and over 3 ft high, eventually was killed by using a gallon (about 6 lb) of sodium chlorate following treatment with copper sulfate crystals. Sucker growth stopped forming, and the thick bark dried up and fell off. In applying treatment to a stump of this type, it is convenient to introduce it between the bark and the wood. Typical formulas for stump killers are shown below. All chemicals in No. 1 to 8 are in powder form.

The material is placed in holes or troughs, then soaked with water to partially dissolve it.

No. 1

| | |
|---|---|
| Potassium nitrate powder | 100 |

No. 2

| | |
|---|---|
| Sodium chlorate powder | 100 |

No. 3

| | |
|---|---|
| Potassium nitrate powder | 90 |
| Copper sulfate powder | 10 |

No. 4

| | |
|---|---|
| Sodium nitrate powder | 90 |
| Ammonium sulfamate powder | 8 |
| Sodium lauryl sulfate powder | 2 |

No. 5

| | |
|---|---|
| Sodium chlorate powder | 90 |
| Copper nitrate powder | 10 |

No. 6

| | |
|---|---|
| Sodium nitrate powder | 60 |
| Potassium dichromate powder | 30 |
| Copper sulfate powder | 10 |

No. 7

| | |
|---|---|
| Sodium nitrate powder | 75 |
| Sodium dichromatic powder | 25 |

No. 8

| | |
|---|---|
| Ammonium persulfate powder | 45 |
| Sodium nitrate powder | 45 |
| Copper sulfate powder | 10 |

No. 9

| | |
|---|---|
| Kerosene | 95 |
| "Sulfonic" N 95 | 3 |
| n-Butanol | 2 |

### Sanitizing Kennels

After thorough cleaning use 1 oz of "Hyamine" 3500 in 2 gal water.

### Mold Control on Walls, Ceilings, etc.

Clean thoroughly, wash well, and spray or sponge with:

| | |
|---|---|
| "Hyamine" 2389 | 1 oz |
| Water | 1 gal |

### Disinfecting Vegetable Sacks

Soak bags for 1 hr in:

| | |
|---|---|
| "Hyamine" 2389 | 1 oz |
| Water | 2 gal |

### Algae Control

The treatment consists of two phases, algaecidal and algistatic. Generally, the addition of 1 gal of a 10% "Hyamine" 2389 solution to each 10,000 gal of water, providing a concentration of 10 ppm, is adequate for algaecidal treatment. This may be added with other water treatment chemicals over a period of several hours to several days. After the algaecidal treatment is complete, dead growth should be removed or the water changed. One gallon of 10% "Hyamine" 2389 solution is added to each 40,000 to 50,000 gallons of make-up water, providing 2-3 ppm "Hyamine" 2389 to prevent further growth. In certain cases, periodic shock treatment may be preferable.

### Laying Hen Feed

| | | |
|---|---|---|
| Ground yellow corn | | 59.81 |
| Soybean oil meal (44%) | | 25.00 |
| Stanbilized lard | | 2.00 |
| Meat and bone meal (50%) | | 2.50 |
| Dehydrated alfalfa meal (17%) | | 2.50 |
| Ground limestone | | 7.25 |
| Dicalcium phosphate (18%) | | 0.50 |
| Salt | | 0.25 |
| Trace mineral premix | | 0.05 |
| Manganese (as sulfate) | 12.20% | |

| | | |
|---|---|---|
| Iron (sulfate) | 5.40% | |
| Copper (sulfate) | 0.73% | |
| Cobalt (sulfate) | 0.20% | |
| Iodine (calcium iodidide) | 0.38% | |
| Zinc (oxide) | 10.00% | |
| Calcium (carbonate) | 5.68% | |
| Vitamin premix | | 0.14 |
| Vitamin A | 4,897 units/lb | |
| Riboflavin | 1.8 mg/lb | |
| Niacin | 25.4 mg/lb | |
| Pantothenic acid | 5.3 mg/lb | |
| Choline | 730.0 mg/lb | |
| Vitamin E | 27.4 mg/lb | |
| Vitamin $B_{12}$ | 0.002 mg/lb | |
| Vitamin D | 225 units/lb | |

### Feed Binder

Poultry and animal feedstuff binder—"Lignosol" FG is used to pelletize poultry and animal feeds. Use is approved, up to 4%, in Canada and the U.S.A. in poultry feeds.

### Salt Block Pressing

Salt-lick and nutritional animal lick-blocks are formulated to contain 1% of "Castorwax." It acts as an internal lubricant during pressure molding as a mold release. In addition it waterproofs the blocks to reduce rain erosion during use.

### Feed Additive for Heifers

Add 10 lb formic acid (90%) to 1 ton unwilted silage. Mix thoroughly. This feed produces more rapid weight gain.

### Sanitizing Kennels

After thorough cleaning use 1 oz of "Hyamine" 3500 in 2 gal water.

### Mold Control for Walls and Ceilings

Clean thoroughly, wash well and spray or sponge with:

| | |
|---|---|
| "Hyamine" 2389 | 1 oz |
| Water | 1 gal |

## CHAPTER IX

# FOODS and BEVERAGES

**Preservatives for Foods or Beverages**

| *Product* | *"Paraben" Concentration* |
|---|---|
| Bakery goods (cakes, pies, pastries, icing, topping, filling) | 0.023–0.045% methyl+0.007–0.015% propyl |
| Cheeses | methyl (dipping) |
| Cider | 0.025–0.04% methyl+0.012–0.017% propyl |
| Creams and pastes | approx. 0.1% total (methyl+propyl) |
| Dried fruits | 0.2% propyl |
| Fish | 0.03–0.06% total (methyl+propyl) |
| Flavor extracts | 0.05–0.1% total (methyl+propyl) |
| Fruit products (fruit salads, juices, sauces, syrups, fillings) | 0.033% methyl–0.017% propyl* |
| Gelatin | 0.05–0.1% methyl |
| Gelatinous coatings | 0.05–0.1% total (methyl+propyl) |
| Jams and jellies | 0.047% methyl+0.023% propyl* |
| Malt extracts | 0.05% total (methyl+propyl) or 0.04% propyl |
| Olives (stuffed with anchovies) | approx. 0.1% total* (methyl+propyl) |
| Pickles | approx. 0.1% total* (methyl+propyl) |

| | |
|---|---|
| Salad dressings | 0.1–0.13% total (methyl–propyl) |
| Soft drinks | 0.02–0.033% methyl–0.01 –0.017% propyl |
| Syrup, chocolate | 0.02–0.03% propyl* |
| Syrups, sucrose | 0.07% methyl or 0.02% propyl |
| Wines | under 0.1% total (methyl & propyl) |

---

**Preservatires for Food Additives**

| | *"Paraben" Concentration* |
|---|---|
| Corn syrup | 0.1% methyl & 0.02% propyl or 0.08% ethyl or 0.04% propyl |
| Gelatin | 0.12% methyl & 0.03% propyl or 0.15% ethyl |
| Gum arabic | 0.12% methyl & 0.03% propyl or 0.1% ethyl or 0.04% propyl |
| Gum trgacanth | 0.12% methyl & 0.03% propyl or 0.1% ethyl or 0.04% propyl |
| Sorbitol | 0.12% methyl & 0.02% propyl |
| Starch | 0.15% methyl & 0.03% propyl |

---

**Cold Sterilization of Beverages**

(Beer, Soft Drinks, Juices, Wines, Cider)

Effective germ control usually requires quantities of "Baycovin" between 5 and 20 g per hectoliter, but quantities as low as 3 to 8 g/hl may suffice for aseptic bottling of thoroughly presterilized beverages as long as other hygienic requirements are strictly observed in the process. For added safety, the additive quantities determined in laboratory tests should be slightly increased in actual operations. It should be mixed thoroughly in beverage just before bottling.

### Increasing Flavor of Food Products

Roasted Nuts

The wet nuts should be coated completely with a mixture composed of:

| | |
|---|---|
| Monosodium glutamate (small crystals) | 1 |
| Corn starch (low density) | 3 |
| Flour salt | 6 |
| Spice as preferred | |

After drying in forced hot air at 425°F, roasting is carried out at 425°F in still air with intermittent tumbling until the nuts have attained the desired color.

| | |
|---|---|
| Popcorn | 1 lb per 25 lb salt |
| Potato chips | 1 lb per 10 lb salt |
| Crackers, all kinds | 1.5 oz per 100 lb |
| Pickles | 1 oz per 100 lb |

Vegetables

| | |
|---|---|
| Asparagus | 3.25 oz per 100 lb |
| Beets | 2.50 |
| Broccoli | 2.50 |
| Brussels sprouts | 2.50 |
| Carrots | 4.00 |
| Cauliflower | 2.50 |
| Corn-creamed style | 2.50 |
| Corn-whole kernel | 3.50 |
| Green beans | 3.25 |
| Lima beans | 3.00 |
| Mixed vegetables | 3.25 |
| Mushrooms | 3.50 |
| Peas | 4.00 |
| Potatoes | 2.00 |
| Spinach | 3.50 |
| Squash | 2.50 |
| Tomatoes | 2.50 |
| Tomato juice | 3.50 |
| Wax beans | 2.50 |

MSG confers another important quality when used in frozen foods: It protects color.

Cheese Products

| | |
|---|---|
| Cheese sauce (such as Welsh rarebit) | 1.5 oz per 20 gal |
| Macaroni and cheese | 1.0 oz per 40 lb |
| Potatoes au gratin | 2-3 oz per 100 lb |
| Grated cheese, Romano or Parmesan | 1-2 oz per 100 lb |

Sea Food

| | |
|---|---|
| Fish, fresh, or for canning or freezing | 2.5 oz per 100 lb |
| Fillets and steaks | 2.5 |
| Shellfish (canned or frozen), shrimp, scallops, oysters, crab, lobster | 2.5 |
| Sardines in tomato paste | 3.5 |
| in oil or mustard sauce | 2.0 |
| Salmon, fresh, or for canning or freezing | 3.0 |
| Tuna | 2.0 |
| Tuna pies, frozen | 2.5 |
| Fish pastes, canned | 4.0 |
| Breaded sea food, frozen (add the MSG to the wet wash) | 2.5 |
| Clam juice cocktails or broth | 3.0 |
| Fish and crab cakes | 3.0 |
| Sea food bisque | 3.5 |
| Sea food salads | 5.0 |

Meat

Fresh cuts of beef, pork, veal, and lamb are greatly improved by the use of MSG. This is especially true of the poorer quality cuts which often do not possess as good a flavor as the choice cuts. Use of MSG by the meat packer on these cuts of meat will insure consumer acceptance. It should be dusted on the meat at the rate of 2.5 oz per 100 lb.

---

**Coffee Whiteners (Synthetic Cream)**

No. 1 (Liquid)

| | Range % (on total wt) | Suggested Levels % (on total wt) |
|---|---|---|
| Fat, Vegetable | 3.0-18.0 | 10.00 |
| Protein (i.e. sodium caseinate) | 1.0-3.0 | 2.00 |
| Corn syrup solids | 1.5-3.0 | 2.50 |

| | | |
|---|---|---|
| Sucrose | 1.0-3.0 | 2.50 |
| "Atmos" 150 | 0.3-0.5 | 0.40 |
| Carrageenan | 0.1-0.2 | 0.20 |
| Stabilizer salts (e.g., sodium citrate) | 0.1-0.3 | 0.15 |
| Flavor | q.s. | q.s. |
| Color | q.s. | q.s. |
| Water | q.s. to 100% | q.s. to 100% |

The dry ingredients are blended with the fat and liquid ingredients and heated to pasteurizing temperature (this temperature varies depending on the particular method used). Once pasteurized the mix is pumped directly to the homogenizer and homogenized at 2,000 to 2,500 total psi (1,500 to 2,000 psi on first stage and 500 psi on second stage) on a two-stage homogenizer, and 1,500 to 2,000 psi on a single-stage homogenizer. It is then cooled to 38°F and stored in a refrigerator.

No. 2 (Powder)

| | |
|---|---|
| Vegetable fat | 35-40 |
| Corn syrup solids (42 D.E.) | 55-60 |
| Sodium caseinate | 4.5-5.5 |
| Dipotassium phosphate | 1.2-1.8 |
| Emulsifier: | |
| a. "Atmos" 150/"Span" 60/"Tween" 60 (60/20/20) | 0.3-0.5 |
| or | |
| b. "Atmos" 150/"Tween" 65 (75/25) | 0.2-0.4 |
| Color | q.s. |
| Flavor | q.s. |
| Anticaking agent | q.s. |

To prepare a powdered whitener, an emulsion concentrate is formed prior to spray drying by dissolving or dispersing the various dry ingredients in just enough water to (1) maintain the solids in solution and (2) impart sufficient fluidity to the concentrate so that it may be pumped. The dissolved solids of the concentrate are usually in the range of 50 to 60%, the higher percentage being strongly recommended. Once the concentrate has been prepared, it is homogenized in such a manner that the fat particles in the dried emulsion will average about 1 micron in diameter. Under normal circumstances this will require about 2,000-2,500 lb total pressure on a two-stage homogenizers. However, it should be remembered that spraying will affect the emulsion, and therefore each manufacturer will have to adjust the pressures to suit his own spraying requirements and the viscosity of the emulsion, also affected by the solids content of the system. At the spray tower, operating conditions should be such that the final product will:

1. Have a moisture content not in excess of 1%.
2. Have a particle size of 125 to 150 microns in diameter.
3. Have entrapped fat globules no larger than 1-3 microns in diameter.

Care should be taken that the heat of crystallization of the fat is removed from the powder prior to packaging, or clumping of the product in the package will result. In some instances, it may be desirable to "instantize" the whitener by agglomerating the powder, making the product more dispersible. The powdered whitener is then packaged in a moisture-proof container.

---

**Canned Puddings**

Chocolate

| | No. 1 | No. 2 | Low-cal. |
|---|---|---|---|
| "Gelcarin" SI | 0.158 | 0.247 | 0.330 |
| "Gelcarin" HMR | — | 0.099 | 0.120 |
| Sugar | 15.385 | 12.465 | — |
| Starch | 2.564 | — | — |
| Salt | 0.039 | 0.049 | 0.060 |
| Fat (including emulsifiers) | 0.986 | 4.056 | — |
| Disodium phosphate | 0.296 | 0.376 | 0.459 |
| Cocoa | 2.564 | 2.672 | 3.139 |
| Artificial sweetener | — | — | 0.130 |
| Fluid whole milk | 78.008 | 80.036 | 95.762 |

Vanilla (or other flavor)

| | No. 3 | No. 4 | Low-cal. |
|---|---|---|---|
| "Gelcarin" SI | 0.140 | 0.228 | 0.300 |
| "Gelcarin" HMR | — | 0.085 | 0.190 |
| Sugar | 10.375 | 11.407 | — |
| Starch | 2.993 | — | — |
| Salt | 0.040 | 0.050 | 0.060 |
| Fat (including emulsifiers) | 2.075 | 8.104 | — |
| Disodium phosphate | 0.279 | 0.380 | 0.420 |
| Cocoa | — | — | — |
| Artificial sweetener | — | — | 0.140 |
| Fluid whole milk | 84.098 | 79.746 | 98.890 |

The carrageenan levels can be varied to obtain a variety of textures and consistencies.

No. 1 and 3 give "spoonable" puddings closely resembling the consistency and texture of a cooked starch pudding with no discernible gel structure. No. 2 and 4 yield puddings with slightly more gel structure than a cooked pudding—between a cooked pudding and a custard in texture, with a creamy delicate body.

---

**Chinese Noodles**

| | |
|---|---|
| Flour (bread) | 10.00 |
| Water | 2.00 |
| Eggs (whole) | 2.00 |
| Salt | 0.19 |

Blend smooth. Roll out thin, allow to dry. Cut in thin strips, dry thoroughly. Fry at 375°F. Noodle dough may be extended through a plate (e.g. spaghetti extruder), then dried and fried or baked.

---

**Butter for Nonwheat-Flour Bread**

| | |
|---|---|
| Cassava flour | 80 |
| Defatted soya flour | 20 |
| Compressed yeast | 2 |
| Salt | 2 |
| Sugar | 2 |
| Water emulsion 1:9 | 10 |
| Glyceryl monostrarate/ water | 60 |

---

**Whipped Topping**

No. 1

| | |
|---|---|
| Fat, m.p. 94-98°F | 30% |
| Sodium caseinate | 2% |
| Sucrose | 7% |
| Corn syrup solids | 3—5% |
| Stabilizer | 0.3—0.5% |
| Emulsifier | 0.35—10.% |
| Flavor and color | q.s. |
| Water | q.s. to 100% |

(Stabilizing salts, such as disodium phosphate and sodium citrate, are used in some instances.)

No. 2

| | |
|---|---|
| Fat (96°F MP) | 30.00 |
| Sodium caseinate | 3.00 |
| Sucrose | 6.00 |
| Corn syrup solids (42 DE) | 2.00 |
| HV CMC, type 7HP | 0.50 |
| Emulsifier* | 0.40-0.90 |

| | |
|---|---|
| Disodium phosphate | 0.05 |
| Color | q.s. |
| Flavor | q.s. |
| Water | q.s. to 100 |

Blend together all dry ingredients (protein, sugar, corn syrup solids, stabilizer, salts) and add water. Mix thoroughly, and add the melted fat and emulsifier. Pasteurize at 160°F for 30 min and homogenize at 1500 total psi (1000 psi first stage, 500 psi second stage). Cool and store overnight at 40°F. Whip until desired overrun and dryness are obtained. This topping will be characterized by good overrun, good dryness, fair to average stiffness and stability. It should be used at once on puddings, gelatin, sodas, and pastry, etc.

---

| | |
|---|---|
| *"Atmos" 150VS/"Tween" 60VS (80/20) | 0.70% |
| or | |
| "Span" 60VS/"Tween" 60VS (60/40) | 0.4% |
| or | |
| "Atmos" 150VS | 0.9% |

No. 3 (Proteinless Whipped Topping)

| | |
|---|---|
| Phase 1 | |
| "Klucel" GP | 0.50 |
| Hot water | 19.50 |
| Phase 2 | |
| Water (cold) | q.s. to 100 |
| Sucrose | 10.00 |
| Salt | 0.80 |
| Lecithin | 0.05 |
| Fat (101°F MP) | 25.00 |
| Emulsifier* | 0.80 |

Phase 1—Disperse "Klucel" powder in hot water at 125°F to 150°F and let stand for at least 5 min to permit the particles to be thoroughly wetted out. "Klucel" will gel if temperature goes below 105°F during this period.

Phase 2—Add cold water (tap water) to lower the temperature of the system to below 95°F. Stir until all "Klucel" particles are completely dissolved. Add sugar and salt.

Melt the fat and emulsifiers together at as low a temperature as possible. Add the hot melt, while stirring, to the cold water containing the dissolved "Klucel", sugar and salt. It is important that the temperature of the fat-in-water mix be held as low as possible to avoid excessive precipitation of the "Klucel". To facilitate this, the temperature of the melted fat should be at the minimum just prior to addition to the "Klucel" solution.

Pasteurize at 160°F for 30 min. Homogenize at 2000 and 500 psi in a two-stage homogenizer and cool as rapidly as possible to 40°F. The mix may then be deaerated (if desired), packed, and frozen for storage.

*"Atmos" 150VS/"Tween" 60VS (60/40)

---

**Bar Fudge Slab**

| | No. 1 | No. 2 | No. 3 | No. 4 |
|---|---|---|---|---|
| Corn syrup (42 DE) | 28.4 | 27.7 | 27.3 | 26.8 |
| Sweetened condensed milk | 16.6 | 16.6 | 16.6 | 16.6 |
| Sucrose | 23.7 | 23.7 | 23.7 | 23.7 |
| Hard fat (92°F) | 3.8 | 3.8 | 3.8 | 3.8 |
| Water | 3.7 | 3.7 | 3.7 | 3.7 |
| Salt | 0.1 | 0.1 | 0.1 | 0.1 |
| Fondant (80-20) | 23.7 | 23.7 | 23.7 | 23.7 |
| "Sorbo" | — | 0.7 | 1.1 | 1.6 |

The corn syrup, water, sweetened condensed milk and "Sorbo" (when used) are warmed on a steam bath and then placed in a steam-jacketed kettle. The dry ingredients are slowly added with agitation and the batch is brought to 246°F. The heat is removed and the fat added. The fondant is added at 203°F and thoroughly incorporated, avoiding excessive agitation. The batches are cast into wooden trays lined with silicone paper and allowed to stand for 24 hr. They are then cut, packaged in confectioner's cellophane, and stored.

---

**Cream Mints, Starch Cast**
(Enrobed)

| | |
|---|---|
| 1—Bob Syrup | |
| Sucrose | 20.0 |
| Corn syrup 42 DE | 11.6 |
| Water | 3.8 |
| "Sorbo" | 0.5 |
| 2—Remelt | |
| Fondant (80/20 242°F cook) | 47.9 |
| Mazetta | 16.0 |
| Invertase | 0.1 |
| Flavor | 0.1 |

Cook bob syrup to 246°F. Cover kettle to wash down sides and insure solution of all sugar crystals. *Remelt*—Cool bob syrup down to 150°F, add fondant and mazetta and mix well. When bob syrup, fondant and mazetta are well mixed, and temperature has been adjusted to 150°F, add flavor and invertase. Pipe into starch molds, let stand 3 hr; dust and enrobe.

---

**Grained Marshmallow**
(Starch Cast)

| | |
|---|---|
| Sucrose | 42.5 |
| Corn syrup 42 DE | 27.2 |
| Gelatin 225 bloom | 9/4.3 |
| Fondant 80/20 | 11.0 |
| "Sorbo" | 1.2 |
| Water | 12.9 |
| Color | q.s. |
| Flavor | q.s. |

The water, sucrose, corn syrup, and "Sorbo" are heated to 180°F (just sufficient heat to dissolve ingredients). The heated solution is examined by refractometer for dissolved solids content and all batches cooked to 79% solids. The solution is then transferred to a Hobart mixer and agitated at low speed. The batch is cooled to 155°F and the fondant thoroughly incorporated. The solution is then agitated at high speed and the geltain-water "solution" is slowly added during agitation. The total batch is whipped to a density of 4 lb per gal. Flavor and color are finally added, and the marshmallow is cast into starch at a temperature of 115 to 120°F.

---

**Canine Confection**
No. 1

| | *lb* |
|---|---|
| Dextrose | 131 |
| Corn syrup (43/42) | 65 |
| Shortening | 15 |
| Milk powder | 99 |
| Salt | 3 |
| Dicalcium phosphate | 75 |
| Activated charcoal | 3 |
| Water | 100 |

Mix together thoroughly and cook to 248°F for caramel or approximately 270°F for hard candy, stirring constantly. After cooling slightly, add and mix in well 9 lb of crude baker's yeast.

Temperatures may vary with degree of pulverization of dicalcium phosphate and yeast. By cooking to 255-275°F and pulling, a kiss type candy may be prepared.

---

**High Cook Caramel, Starch Cast**

| | |
|---|---|
| Sucrose | 19.3 |
| Corn syrup 63 DE | 39.9 |
| Sweentened condensed milk | 29.6 |
| Butter | 3.5 |
| Salt | 0.7 |
| Caramel paste | 3.3 |
| Lecithin | 0.1 |
| Invert sugar | 3.6 |
| "Sorbo" | — |

Caramel Paste

| | |
|---|---|
| Corn syrup 63 DE | 28.8 |
| Sucrose | 24.7 |
| Sweetened condensed milk | 11.8 |
| Starch, 60 fluidity | 9.0 |
| Vegetable oil | |

| | |
|---|---|
| (76° m.p.) | 6.0 |
| Butter | 1.3 |
| Vanilla | 0.1 |
| Water | 18.3 |

All ingredients are weighed into a stainless steel bowl and heated with steam to 150°F, using agitation to keep the starch and condensed milk from sticking and burning. The paste is held at 150°F and checked to be sure it is completely homogeneous. Heat is again applied and the paste cooked to 22·°F, cooled, and held for the caramel.

Caramel

All ingredients are weighed into a kettle and heat gradually applied to warm batch slowly to 150°F. Again care is taken to prevent sticking or scorching. When the batch is homogeneous, and at 150°F, the batch is cooked to 255°F. The batch is cooled to 245°F and maintained at this temperature while being cast into starch (90°F). After 24 hours, the caramels are removed from starch, dusted and enrobed with dark sweet chocolate. The samples are then boxed and wrapped in confectioner's cellophane.

No. 2

| | *lb* |
|---|---|
| Gelatin (150 bloom) | 8.0 |
| Water | 50.0 |
| Water | 80.0 |
| Dextrose | 161.0 |
| "Amaizo" corn syrup (43/42) | 136.0 |
| Powdered blood | 50.0 |
| Di-Calcium phosphate | 162.5 |
| Activated charcoal | 6.5 |
| Salt | 6.5 |
| Crude yeast | 19.5 |
| Bone meal | 50.0 |
| Liver | 50.0 |

Soak gelatin in water. Boil to 220°F, cool to 150°F, and add gelatin solution. Stir to dissolve. Add to the beater and beat until 1 gal wt=3½ lb. Blend together—add gradually and mix in well.

Quantities of Blood powder, Bone meal and Liver meal may be varied to control basis.

No. 3

| | |
|---|---|
| Water | 50 |
| Gelatin | 8 |
| Water | 30 |
| Dextrose | 100 |
| Corn syrup | 157 |
| Dextrose | 85 |
| Water | 50 |
| Dicalcium phosphate | 125 |
| Activated charcoal | 5 |
| Salt | 5 |
| Crude baker's yeast | 15 |

Heat to dissolve gelatin (120°F). Mix together in bowl, start beating add warm gelatin solution and continue beating. Boil to 240°F and add slowly to (2), beat until light. Add and beat until one gal weighs 3½ lbs.

Any of the following flavors may be added:

1½ lb dry-soup flavor.
1½ lb yeast extract.
1½ oz liverwurst flavor.
1½ oz sardine flavor.
45 lb cocoa.
4 lb bouillon flavor.

### Mince Meat, Dry

| | |
|---|---|
| Apples (dry) | 50.00 |
| Brown sugar | 40.00 |
| Seedless raisins | 100.00 |
| Beef suet (ground) | 15.00 |
| Meat (cooked and ground) | 60.00 |
| Spices | as desired |
| Flavor | as desired |
| Sodium benzoate | 0.25 |
| "Gelex" starch | 10.00 |

Blend together thoroughly. Add corn syrup 43 blend in well and press into cakes.

For use this mince meat is combined with the proper amount of boiling water, additional flavor is added and the filling added to unbaked pie shells. Somewhat longer cooking is required to endure tenderness.

### Chili Con Carne

| | |
|---|---|
| Lean beef chucks through ¼″ plate | 65.00 |
| Beef suet through ½″ plate | 35.00 |
| Dehydrated chopped onions | 0.75 |
| Canned Green Chiles | 2.00 |
| Tomato Puree | 4.00 |
| Milk (powdered) | 5.00 |
| Salt | 2.50 |
| Corn syrup | 1.50 |
| Chili Powder | 4.00 |
| Paprika | 0.75 |

Place suet in jacketed kettle and heat until liquid. Add beef and cook 2 to 2½ hr or until most of the moisture has evaporated; add all other material, including the powdered milk, after first hour of cooking.

### Tartar Sauce

| | |
|---|---|
| "Kelcoloid" HVF | 3.26 |
| Water | 180.00 |
| Sweet pickle relish* | 132.74 |

| | |
|---|---|
| Vegetable oil | 117.94 |
| Spirit vinegar (100 gr.) | 76.28 |
| Sugar | 31.40 |
| Salt | 24.74 |
| Fresh whole eggs | 23.00 |
| Chopped capers | 6.60 |
| Dry mustard | 3.84 |
| Chopped parsley, dry | 0.20 |

Pre-mix "Kelcoloid" HVF with 12 grams sugar until uniform. (If "Kelcoloid" DH is used, there is no need for a sugar pre-mix.) Dissolve pre-mix in 75-80°F water with strong agitation, 8-10 min. While still mixing, add eggs and mix 1 min. While mixing add vinegar, sugar, mustard and salt and mix 1 min. While mixing add vegetable oil slowly and mix 1 min. Homogenize at 1000 psi (if desired). Combine homogenized mixture with parsley, capers, and sweet relish and mix 2 min slowly. Fill at room temperature with mechanical vacuum at 10-14 in.

*If sweet picke relish is not stabilized with "Kelcoloid" DH to give 100% drained weight, use 3.725 g "Kelcoloid" HVF (0.6208%) instead of 3.26 g per batch.

### Sweet Pickle Relish

| | |
|---|---|
| "Kelcoloid" | 9.5 g |
| Water | 12.0 fl oz |
| Chopped, desalted, pressed pickles | 5.0 lb |
| Sugar | 3.0 lb |
| Vinegar (250 grain) | 12.0 fl oz |
| Mustard seed | 25.0 g |
| Celery seed | 3.0 g |
| Chopped red peppers (dehydrated) | 10.0 g |
| Relish oil | to taste (0.5 g) |
| | approx. 10.0 lb |

### Hamburger Relish

| | |
|---|---|
| "Kelcoloid" | 10.5 g |
| Chopped, desalted, pressed pickles | 5.0 lb |
| Vinegar (250 grain) | 12.0 fl oz |
| Fresh chopped onions | 200.0 g |
| Chopped red peppers (dehydrated) | 10.0 g |
| Sugar | 3.0 lb |

| | |
|---|---|
| Mustard seed | 40.0 lb |
| Celery seed | 2.8 lb |
| Relish oil | to taste (0.5 g) |
| Water | 14.0 fl oz |
| Tomato paste (26%) | 100.0-150.0 g |
| Soluble paprika | (20,000 W) to desired level |
| | approx. 11.0 lb |

## Green Tomato Relish

| | |
|---|---|
| "Kelcoloid" | 10.8 g |
| Water | 8.0 fl oz |
| Chopped pressed green tomatoes | 5.0 lb |
| Salt | 70.0 g |
| Sugar | 3.5 lb |
| Vinegar (250 grain) | 12.0 fl oz |
| Mustard seed | 30.0 g |
| Celery seed | 3.4 g |
| Soluble pepper (white) | 0.8 g |
| Chopped red peppers (dehydrated) | 15.0 g |
| Chopped onions (fresh) | 250.0 g |
| Relish oil | to taste (0.6 g) |
| | approx. 12.5 lb |

## Spice Tea

| | |
|---|---|
| Allspice | 4.25 oz |
| Celery seed | 25.00 |
| Parsley seed | 22.50 |
| Sweet basil | 3.50 |
| Thyme | 5.25 |

Place spices in 5 gallons of cold water for 5 minutes and agitate, while simmering for 15 minutes.

Strain and add enough water to make 5 gal.

CHAPTER X

# INKS

### Mimeograph Ink Vehicle

| | |
|---|---|
| Mineral oil | 28 |
| Lanolin | 24 |
| Sugarcane wax | 23 |
| "Surfactol" 318 | 20 |
| Beeswax | 5 |

### Indelible Transfer Ink

| | |
|---|---|
| "Amberol" | 150 |
| "Castorwax" | 45 |
| Carnauba wax | 39 |
| Ozokerite | 32 |
| No. 6 litho varnish | 35 |
| 15 Oil | 60 |
| Cobalt drier | 1 |
| Butyl carbitol | 10 |
| Cadmium red | 215 |

### Antitoning Offset Ink

The addition of 1% "Aerosil" R-972 to an offset ink creates a barrier between aqueous and oil phases.

### Fluorescent Letterpress Ink

| | |
|---|---|
| "Alvco" 730-S | 23.0 |
| "Versatyl" vehicle XLV-4 | 18.0 |
| "Magle" oil 535 | 8.0 |
| W-370 Poly compound | 5.0 |
| P-6000-G "Velva-Glo" Fluorescent pigment | 45.0 |

Mill two passes on three-roller mill, then add driers, 1.0 (0.5 lead naphthenate 24% Pb, 0.5 manganese naphthenate 6% Mn).

### Fluorescent Offset Printing Ink

No. 1

| | |
|---|---|
| "Alvco" 730-S | 18.20 |
| "Alvco" 2517 | 9.00 |
| "Scope" resin 142 | 10.80 |
| "Zonyl" S-13 | 0.13 |
| "Magie" oil 535 | 10.80 |
| P-6000-G "Velva-Glo" Fluorescent pigment | 50.00 |

Mill two passes on three-roller mill, then add driers, 1.0 (0.5 lead naphthenate, 24% Pb, 0.5 manganese naphthenate 6% Mn).

No. 2

| | |
|---|---|
| "Alvco" 730-S | 12.0 |

| | |
|---|---|
| "Versatyl" vehicle XLV-3 | 30.0 |
| "Magie" oil 535 | 8.0 |
| W-370 Poly compound | 5.0 |
| P-6000-G "Velva-Glo" Fluorescent pigment | 44.0 |

Mill two passes on three-roller mill, then add driers, 1.0 (0.5 lead naphthenate, 24% Pb, 0.5 manganese naphthenate 6% Mn).

**Fluorescent Web Offset Heat-Set Ink**

| | |
|---|---|
| XLH 403 | 50.0 |
| XLV-4 | 8.0 |
| "Magie" oil 470 | 10.0 |
| P-6000-G "Velva-Glo" Fluorescent pigment | 50.0 |

Mill two passes on three-roller mill.

---

**Gravure Printing Ink**

| | No. 1 | No. 2 | No. 3 |
|---|---|---|---|
| Phthalo blue G pigment (dry) | 12.0 | — | — |
| "Polycin" 70D dispersion "A" | — | 30.0 | — |
| Dispersed paste "F"* | — | — | 52.2 |
| "Parlon" S-5 50% in toluol | 50.0 | 50.0 | 29.2 |
| Dispersing agent (soya lecithin) | 1.0 | — | — |
| Dioctyl phthalate | 7.5 | — | 7.5 |
| Toluol | 29.5 | 20.0 | 11.1 |

*The composition of dispersed paste "F" is:

| | |
|---|---|
| Phthalo blue G pigment | 23 |
| "Parlon" S-20 | 20 |
| Toluol | 57 |

Ink No. 1 is a ball-mill grind which shows the matte, hazy finish typical of ball-milled phthalocyanine. Ink No. 2 is a propeller-mixer reduction of the three-roller grind dispersion in "Polycin" 70D. This ink has excellent gloss and transparency, especially for such a hard-grinding pigment as phthalo blue G. Ink No. 3 is a propeller-mixer reduction of a commerically dispersed phthalo blue G paste.

| | No. 4 | No. 5 | No. 6 |
|---|---|---|---|
| Phthalo blue G (dry) | 12.0 | — | — |
| "Polycin" 70D dispersion "A" | — | 30.0 | — |
| Chip "D"* | — | — | 20.0 |
| Dispersing agent (soya lecithin) | 1.0 | — | — |
| Dioctyl phthalate | 15.0 | — | 13.0 |
| Nitrocellulose ($\frac{1}{4}$ sec. R.S.) 35% in ethyl acetate | 44.0 | 44.0 | 26.9 |

| Ethyl acetate | 28.0 | 26.0 | 40.1 |
|---|---|---|---|

| *The composition of chip "D" is: | |
|---|---|
| Phthalocyanine blue GS pigment | 60 |
| ¼ sec. R.S. nitrocellulose | 30 |
| "Santicizer" B-16 | 10 |

## Silk Screen Ink

| | No. 1 | No. 2 |
|---|---|---|
| EHE-low | 5.8 | 6.0 |
| "Cellolyn" 102 | 3.5 | 8.0 |
| "Cellolyn" 95-80T | — | 11.0 |
| "Pentalyn" K | 7.1 | — |
| "Hercolyn" D | 8.1 | — |
| Pigment | 44.3 | 44.0 |
| Solvent | 31.2 | 31.0 |

These formulas may be used with any of the following solvent blends:

| | | | |
|---|---|---|---|
| "Atlantic" 49 | 85 | 75 | — |
| "Solvesso" 100 | 12 | 12 | 48 |
| "Sun" spirits | — | 10 | 42 |
| n-Propanol | 3 | 3 | — |
| n-Butanol | — | — | 10 |

## Marking Crayon

| | |
|---|---|
| "Castorwax" | 18 |
| Ceresin | 40 |
| Carnauba wax | 17 |
| Paraffin wax | 20 |
| Beeswax | 5 |
| Talc | 50 |
| Pigment | 15 |

Chapter XI

# LEATHER

## Deliming Calfskin

(Basis: Scudded weight)

| *Operation* | *Material* | *Time, hr* |
|---|---|---|
| Wash | Add skins to 416% water at 75°F in a paddle. | 2 |
| Delime | Drain and refill with 416% water at 75°F. Add 0.5% ammonium chloride. | 1 |
| | Slowly add 0.5% of 90% formic acid in 8% water. Proceed with addition of bating enzyme. | 1 |

## Pretanning Chamois Leather

(Basis: Drained pickled skins)

| *Operation* | *Material* | *Time, hr* |
|---|---|---|
| Washing | Add 200% of water, and adjust pH to about 2.3 with sulfuric acid | 0.5 |
| Tanning | Drain to 100% water, and add 1.5% of "Ucar" tanning agent G-50 | 0.5 |
| pH Adjustment | Add 2% of sodium formate to adjust pH slowly to about 4 | 2.0 |
| Neutralize | Add 1.5 to 3% of sodium bicarbonate to raise pH to about 7 | 2-3 |
| Wash | Wash with running water and proceed to normal cod oil tannage | 0.5 |

## Retannage of Mellow Hand-Sewn Moccasin Leather

| *Operation* | *Material* | *Time, hr* |
|---|---|---|
| Wash | Use 200 to 250% of water at 130°F | 0.1 |

| | | |
|---|---|---|
| | (55°C). | |
| Neutralize | Drain to 100% float. Adjust pH to 3.0-3.5. | 0.1 |
| Retan | Add 3 to 5% of 10% commercial chrome tanning solution (33% basic). | 0.2 |
| | Add 5 to 7.5% of "Ucar" tanning agent G-50 in two feeds 10 min apart. | 0.8 |
| Neutralize | Add sodium acetate or sodium formate to pH of 4 to 5. | 0.3 |
| Wash | Drain and wash. Drain to 150% float. Dye and fat-liquor to desred properties. | 0.2 |

**Tanning Calfskin for Side Upper Leather**

(Basis: Scudded weight)

| *Operation* | *Material* | *Time, hr* |
|---|---|---|
| Neutralize | Add 150% of water and 6% of salt at 68°F (20°C). Adjust pH to about 3.5 with 0.5 to 1.0% sulfuric acid. | 1-2 |
| Tannage | Drain off about 1/3 of pickle liquor, add 1.5 to 2% of a 14% chrome solution (33% basicity). | 2.0 |
| | Second feed of 1.5 to 2% of chrome tanning agent with 5 to 7.5% of "Ucar" tanning agent G-50 (two feeds 15 min apart). | 4.0 |
| Neutralize | Wash with running water at 122°F (50°C); add about 2% of borax to raise pH to about 5.5. | 0.5 |

**Formaldehyde Tannage**

(Basis: Drained, degreased, pickled sheepskins)

| *Operation* | *Material* | *Time, hr* |
|---|---|---|
| pH Adjustment | Add 6% of salt to 100% of water at 75°F. | 0.2 |
| | Adjust to pH 4.5 with 2 to 4% of sodium acetate. | 0.3 |
| Tanning | Add 6% of formaldehyde (37% aqueous). Let drum stand overnight. | 4.5 |
| Neutralize and Set | Add 50% of water, and steam to about 100°F (38°C) and 200% float. | 0.5 |

| | | |
|---|---|---|
| Tannage | Adjust to pH 9 with 1 to 2% of sodium carbonate. | 1-2 |
| Wash | Wash well with running water to remove formaldehyde. Proceed with coloring and fat-liquoring. | 0.5 |

---

**Simultaneous Glutaraldehyde/Chrome Tannage of Shearling**

(Basis: Pickled, scoured shearlings)

| *Operation* | *Material* | *Time, hr* |
|---|---|---|
| Wash | Add 500% of water. Drain to about 200% float. | 0.5 |
| Buffer | Add 6% sodium formate or sodium acetate to adjust pH to 3.5 to 4.0. | |
| Tanning | Add 2.5% of "Ucar" tanning agent G-50. pH about 3.5; temperature 80°F. | 0.5 |
| | Add 2.5% of "Ucar" tanning agent G-50. pH about 3.5; temperature 80°F. | 1.0 |
| | Add 4% of commercial chrome solution (33% basic); pH about 3.5. | 1.0 |
| Neutralize | Add 1 to 2% of sodium bicarbonate to adjust pH to 6-7. | 2.0 |
| Wash | Make two washes, and then horse leather for next step in process. | 0.3 |

---

**White Side Leather Retannage**

(Basis: Blue shaved weight)

| *Operation* | *Material* | *Time, hr* |
|---|---|---|
| Wash | Water at 80°F. | 0.1 |
| Neutralize | Drain and add 150% of water at 90°F (32°C). Adjust pH to about 2.5 with 0.6% of sulfuric acid. | 0.1 |
| Bleach | Add 1.5% of oxalic acid and 6% of naphthalene sulfonic acid syntan (pH about 2.5). | 0.5 |
| Wash | Running water at 100F. | 0.2 |
| | Running water at 130°F. | 0.1 |
| | Drain and float at 150% at 130F. | |
| Retannage | Add 4% of "Ucar" tanning agent G-50 (two feeds 15 min apart) (liquor pH 3.5). | 1.5 |

| | | |
|---|---|---|
| Wash | Add water at 130°F. Continue process with normal pigmentation procedure. | 0.1 |

---

**Plumping Chrome-tanned Calf Skins**

Wash 20 min at 100°F. Neutralize with ½% sodium bicarbonate for 20 min at 80°F. Wash 10 min at 80°F and 10 min at 100°F. Add: 3% sumac extract, ¾% "Chembark" T, 3% "Leukanol" ND, 4/10% "Gantrez" AN-139. Run 30 min at 120°F. Add: % potassium oxalate, 1% formic acid. Run 15 min. Wash 10 min at 120°F.

To upgrade side leather by imparting body and roundness, "Gantrez" AN is usually applied to the blue cream leather in a drum using the dry powder and common salt in a short float.

---

**Dyeing Suede Upper Leather**

(Basis: Dried buffed calf)

| *Operation* | *Material* | *Time, hr* |
|---|---|---|
| Wetting | Add 500% of water at 122°F (50°C) and 1 lb of "Tergitol" Nonionic 15-S-9. | 1-2 |
| Dyeing | Drain and add 150% of water at 131°F (55°C) and 1 lb of ammonia. | 0.2 |
| | Add 3% of acid type dye (dissolved). | |
| | Add 3% of acid type dye (dissolved). | 0.5 |
| | Second addition of 3% of dye. | 0.5 |
| Setting Dye | Add 3.5% of formic acid, 90%. | 0.5 |
| Washing | Wash with water, remove from drum, and horse up. Proceed to drying. | 0.2 |

---

**Leather Garment Treatment**

| | |
|---|---|
| "Hoechst" Wax OM | 1.5 |
| "Hoechst" Wax RT | 0.5 |
| Slab paraffin 52/54° | 3.0 |
| Benzine (100 to 140°C) | 65.0 |
| Methylene chloride | 30.0 |

---

**Leather Processing**

(Fatliquoring)

Stable emulsions of cod, neatsfoot, and mineral oil are prepared by using "Surfactol" 365 at a level of 10% based on the weight of oil. Such emulsions are stable in the presence of the salts and acids remaining in the leather after prior processing steps. The "Surfactol" 365 promotes the even penetration of these oils into the leather and prevents mottling.

**Degreaser for Pickled Leather**

| | |
|---|---|
| "Surfactol" 318 | 2 |
| "Surfactol" 365 | 2 |
| Kerosene | 40 |
| Water | 56 |

Any degreasing solution remaining in the leather after this operation will have no adverse effect on further processing of the skins.

CHAPTER XII

# LUBRICANTS

## Metal Working Fluid

(Cutting oil)

| | No. 1 | No. 2 | No. 3 |
|---|---|---|---|
| "Surfactol" 318 | — | 18.00 | 12.50 |
| "Surfactol" 340 | 7.50 | 18.00 | 12.50 |
| "Surfactol" 365 | 42.50 | — | — |
| Tertiary nonyl polysulfide | — | 5.00 | — |
| Petroleum sulphonate | — | 5.00 | 5.00 |
| Dipropylene glycol | 50.00 | 54.00 | 44.83 |
| Dye | — | 0.02 | — |
| Masking agent | — | 0.15 | — |
| Water | 150.00 | 150.00 | 600.00 |
| Hexylene glycol | — | — | 20.00 |
| Dibenzyl disulfide | — | — | 2.00 |
| Polyoxyethylene octadecyl amine | — | — | 2.00 |
| Phenyl mercuric acetate (18% solution) | — | — | 1.00 |

All parts by weight. The base grinding fluid is prepared by mixing the two "Surfactols" with the dipropylene glycol and then forming an emulsion of this nonaqueous mixture in water.

No. 4

| | |
|---|---|
| "Castorwax" | 10.70 |
| Stearic acid | 2.14 |
| Triethanolamine | 1.07 |
| "Triton" X-100 | 0.53 |
| Water | 85.56 |

Melt the "Castorwax" and stearic acid together and bring to a temperature of about 212°F. At the same time add the triethanolamine and "Triton" X-100 to about one third of the total water and bring to boiling. Add the melted "Castorwax"-stearic acid mix slowly to the TEA-water solution with rapid agitation. While the temperature is still above 200°F, run the emulsion through a colloid mill several times to ensure a small particle size. The remaining water at about 200°F can then be added to the emulsion in the colloid mill. Slowly cool the mixture to room temperature with continued agitation.

No. 5

"Castorwax" can also be emulsified with cationic emulsifying agents to give emulsions which are stable in acid solutions.

| | |
|---|---|
| "Castorwax" | 21.0 |
| "Triton" X-100 | 1.1 |
| "Aerosol" C-61 | 2.6 |
| Water | 75.3 |

Add "Triton" X-100 to one third of the water and heat to 212°F. Add "Aerosol" C-61 to "Castorwax" and heat to 212°F. When the "Castorwax"-"Aerosol" C-61 mix begins to foam or boil, remove from the heat and add slowly to the water with very rapid agitation. The temperature of the water must be maintained above 200°F during the addition of the "Castorwax" so that the temperature, after all the "Castorwax" has been added, remains above 200°F. Continue stirring, allowing the temperature to fall slowly to approximately 170°F. When the temperature has fallen to 170°F the remaining two-thirds of the water can be worked in slowly, with moderate agitation. Emulsions of castorwax containing cationic emulsifiers tend to thicken as they cool after all the water has been added. Homogenize or violently agitate the batch after it is thoroughly cooled, to produce the final low viscosity emulsion.

No. 6

| | |
|---|---|
| "Ultraphos" 41 | 10-15 |
| Alkanolamine | 3-5 |
| Sodium nitrite (if desired) | 4-8 |
| Base oil | balance |

Alkanolamine may be triethanolamine, diethanolamine, or aliphatic polyamines. The base oil may be a 100-sec Saybolt viscosity naphthenic type from any major oil company. A uniform product can be obtained with simple mixing and warming to 120-125°F if necessary. The concentrate is emulsified in 10-20 volumes of water for use.

---

**Lubrication Grease**

| | |
|---|---|
| Paraffinic oil | 93.9 |
| "CAB-O-SIL" H-5 | 5.1 |
| "Tween" 61 | 1.0 |

The greases are made by premixing all ingredients at 250 to 300°F with a propeller stirrer and then milling on a three-roll mill. Greases are usually given two passes through the mill—the first a quick "open" pass and the second at a tight setting. In both passes, the clearance between the last two rolls is set at 0.001 in. To prevent moisture condensing on the rolls and introducing water into the greases, hot water at 140°F is circulated through the rolls. Most of the work is performed with a highly refined paraffinic oil with a viscosity of 2000 SSU at 100°F and a viscosity index of 100.

---

**Chain Belt Lubricant**

Viscous lubricating solutions are used on belt conveyors carrying bottles in beverage plants. Low foam, absence of rusting, and lubricity are required. The following concentrate can be diluted by the operator and dripped onto the belt:

| | |
|---|---|
| "Varion" 1084 | 30 |
| Water | 70 |

---

**Rubber and Resin Lubricant**

"Surfactol" 365, diluted with water or alcohol, can be used as an automotive rubber lubricant. Dissolved in water, it can also be used as a spray lubricant and mold-release for polystyrene and phenolic resins.

---

**Packing Leather Lubricant**
(Water- and Grease-Resistant)

| | |
|---|---|
| "Castorwax | 30 |
| Baker's 40 Oil or Vorite" 125 | 70 |
| "Paricin" 220 | 35 |
| "Paricin" 8 | 32 |
| "Castorwax" | 25 |
| Baker's 30 Oil or "Vorite" 120 | 75 |
| Graphite | 5 |

Mineral Fillers (asbestos, etc.) to desired consistency.

---

**Aerosol Lubricant**

| | |
|---|---|
| Molybdenum disulphide Lubricant | 50.0 |
| Naphtha | 50.0 |
| Methylene chloride | 40.0 |

Aerosol Formulation

| | |
|---|---|
| Concentrate | 50.0 |
| "Genetron" 12/11 (30:70) | 50.0 |

The ingredients in the concentrate are mixed at room temperature. The formulation can be packaged in an unlined, side-seam container fitted with a standard valve and spray actuator. Shake well before using.

---

**Powdered Metals Lubricant**

One-half percent of powdered "Castorwax" or powdered "Paricin" 285 added to powdered metals, such as tungsten carbide, brass, zinc, lead, iron, etc., gives good lubrication in forming metal parts.

---

**Sintered Ceramic Lubricant**

"Castorwax" used up to 2% with powdered ceramic materials gives excellent lubrication during molding, increased strength and impact resistance to the finished item and on firing burns away cleanly with no ash residue.

---

**Dry High Temperature Lubricant**

| | |
|---|---|
| Lead monoxide | 95 |
| "Cab-O-Sil" | 5 |

---

**Radiation-Proof Bearing Lubricant**

| | |
|---|---|
| "K" sodium silicade | 22 |
| Molybdenum disulfide | 71 |
| Graphite | 7 |

CHAPTER XIII

# METALS

### Etchants for Printed Circuits

No. 1

| | |
|---|---|
| Cupric chloride | 38-40 |
| Water | 62-60 |

The advantage of cupric-chloride-based etchants is that of relatively easy regeneration when used to etch copper. The oxidation of cuprous copper to cupric can be done electrically.

No. 2

| | |
|---|---|
| Ferric chloride | 38.8-39.8 |
| Water | 60.9-59.6 |
| Hydrochloric acid, 31.5% | 0.3-0.6 |

When used as an etchant for copper, ferric chloride presents a problem in regeneration. The ferrous chloride content of the fresh etchant should be as low as is practical since ferrous chloride is the degradation product. Some etchers find that 35-36% ferric chloride is more effective. Operating temperatures range from 110 to 120°F.

No. 3

| | |
|---|---|
| Ammonium persulfate | 25 |
| Water | 75 |

This etchant does not have good keeping properties and must be made up shortly before use.

No. 4

The chromic-acid-sulfuric acid based etchants find extensive use at the present time. They are prepared in a variety of formulations. Either sodium dichromate and sulfuric acid can be used as the source of chromic acid, or technical chromic acid flakes can be used. In the case where the chromic acid is generated in place, the resulting sodium sulfate tends to get in the way and reduce the capacity of the etchant. However, this type of etchant is somewhat lower in cost. There is an advantage in the use of chromic acid in that it yields a product having high capacity—etching machines can be run longer before the etchant has to be replaced. Sodium chloride can be used as a catalyst if used fresh. On standing, sodium-chloride-catalyzed etachants tend to "gas." Where ammonium chloride is used, it overcomes this tendency to a great extent. Relatively few formulations of this type represent the ideal. Regeneration of the chromic-acid sulfuric-acid etchants poses a serious problem. While the spent etchant may contain a variety of essentially valuable metals such as cobalt and nickel, it presents a problem in recovery of these materials on a small scale. The investment in equipment necessary to do the job overshadows the potential return. However, it is a shame to have to dispose of the spent etchant as waste material.

Etchant formulations are of two types. General purpose etchants are manufactured in good volume. On the other hand, an equally good volume is made to meet specific requirements. Each manufacturer considers his formulations as top secret. The following formulas will serve as guides to this type of etchant.

No. 4a

| | |
|---|---|
| Water | 1154 |
| Catalyst | 16 |
| Sodium dichromate | 750 |
| Sulfuric acid (93.19%) | 1430 |

No. 4b

| | |
|---|---|
| Water | 1102 |
| Catalyst | 48 |
| Sodium dichromate | 750 |
| Sulfuric acid (93.19%) | 1450 |

No. 5

| | |
|---|---|
| Water | 954 |
| Catalyst | 16 |
| Sodium dichromate | 750 |
| Sulfuric acid (93.19%) | 1630 |

No. 6

| | |
|---|---|
| Water | 902 |
| Catalyst | 48 |
| Sodium dichromate | 750 |
| Sulfuric acid (93.19%) | 1650 |

No. 7

| | |
|---|---|
| Water | 870 |
| Catalyst | 80 |
| Sodium dichromate | 750 |
| Sulfuric acid (93.19%) . | 1650 |

No. 8

| | |
|---|---|
| Water | 2335 |
| Catalyst | 20 |
| Chromic acid | 385 |
| Sulfuric acid (93.19%) | 610 |

No. 9

| | |
|---|---|
| Water | 2295 |

| | |
|---|---|
| Catalyst | 20 |
| Chromic acid | 400 |
| Sulfuric acid (93.19%) | 635 |

No. 10

| | |
|---|---|
| Water | 2215 |
| Catalyst | 20 |
| Chromic acid | 385 |
| Sulfuric acid (93.19%) | 730 |

No. 11

| | |
|---|---|
| Water | 2068 |
| Catalyst | 20 |
| Chromic acid | 400 |
| Sulfuric acid (93.19%) | 762 |

No. 12

| | |
|---|---|
| Water | 2005 |
| Catalyst | 20 |
| Chromic acid | 392 |
| Sulfuric acid (93.19%) | 933 |

In preparing etchants employing sulfuric acid, it is beneficial to dilute the concentrated acid with at least part of the water first, followed by a standing period to dissipate the appreciable heat of dilution. Or the concentrated acid can be replaced with an equivalent amount of 50% or 60% acid. The sulfuric acid should be added last, after the other materials are in solution.

The formulas shown are typical of this type of product. The theoretical amount of sulfuric acid, required to utilize the chromic acid present, will be seen to range from about 50% of the calculated amount to about 75% of this figure. Three moles of sulfuric acid are required per mole of chromic acid anhydride. However, if the theoretical amount of sulfuric acid is used, the etchant tends to run wild. Where anything above 75% theoretical sulfuric acid is used, the etchant must be tailor-made to fit a specific etching job.

---

**Aluminum Etching**

Sodium gluconate is used in etching, providing a smooth mirror-like finish instead of the satin finish usually attained. A suggested bath consists of 25 lb caustic soda and ¾ lb sodium gluconate per 100 gal of bath, at a temperature of 150 to 160°F.

---

**Noncorrosive Aluminum Cleaner**

No. 1

| | |
|---|---|
| Soda ash | 13.0 |
| Tripolyphosphate | 18.0 |
| Sodium acid pyrophosphate | 8.5 |
| Sodium silicate | 7.0 |
| Trisodium phosphate | 16.0 |
| Sodium metasilicate pentahydrate | 32.0 |
| "Neutronyx" 600 | 3.0 |
| "Maprofix" powder LK | 2.5 |

Use from 2 to 3 oz per gal for ultrasonic cleaning and from 4 to 6 oz for soak cleaning.

No. 2

| | |
|---|---|
| Sodium bicarbonate | 65.0 |
| Sodium sulfate | 15.0 |
| Borax | 12.0 |
| Tripolyphosphate | 2.0 |
| "Xynomine" powder No. 20 | 4.0 |
| "Maprofix" LK | 2.0 |

Use a dilution of 2 to 4 oz per gal or 6 oz per gal for heavy soils at 170°F.

**Aluminum Brighter and Metal Conditioner**

This product is effective in removing varnish and oxides from aluminum and metal surfaces.

| | |
|---|---|
| Phosphoric acid (85%) | 45.0 |
| "Triton" X-102 | 12.0 |
| Butyl "Cellosolve" | 25.0 |
| Orthodichlorobenzene | 5.0 |
| Water | 13.0 |

Add phosphoric acid to water first; then add "Triton" X-102 and butyl "Cellosolve," leaving the orthodichlorobenzene for last. Dilute with 3 parts of water for use.

**Liquid Metal Cleaner**

| | |
|---|---|
| "Petro" BA or "Petro" 11 Liquid | 10 |
| Sodium metasilicate pentahydrate | 20 |
| NTA $Na_3$ pentahydrate | 1 or 2 |
| Nonionic (nonyl phenol or linear alcohol ethoxylate) | 1 or 2 |
| $H_2O$ | 66 |

At a concentration of 10% or above, the "Petro" 11 prevents the etching and darkening of aluminum. Recommended use: 1 to 2 oz per gal of water. Recommended for general hard-surface cleaner of floors, walls and metal surfaces. Not recommended for automobile exteriors or where streaking of the surface is objectionable.

**Copper Cleaners**

| | | | |
|---|---|---|---|
| A. | "Veegum" K | 2.07 | 2.07 |
| | "Kelzan" | 0.23 | 0.23 |
| | Water | 78.55 | 78.55 |
| B. | "Snow Floss" | 13.60 | 13.60 |
| C. | Buffer solution* | q.s. | q.s. |
| D. | "Triton" X-102 | 4.65 | 4.65 |
| | Benzotriazole | 0.90 | — |
| | Lauryl thioglycolate | — | 0.90 |
| | Preservative | q.s. | q.s. |

| | | |
|---|---|---|
| Perfume | q.s. | q.s. |
| Perfume | q.s. | q.s. |
| Color | q.s. | q.s. |

*Buffer solution: 1 part 1M $H_3PO_4$,
1.46 parts 125 g/liter $Na_3PO_4$.

1. Dry blend "Veegum" and "Kelzan" and add to the water slowly, agitating continually until smooth.
2. Add B to I gradually and mix until smooth, then buffer this mixture with buffer solution C to a pH of 2.5.
3. Mix components in D until a clear solution is formed. Special care should be taken to avoid incorporation of air.
4. Add 3 to 2 very slowly and mix until uniform.

Apply copper cleaner with damp cloth. Rinse and dry. Polish with a clean dry cloth.

**Aerosol Oven Cleaner**

| | | |
|---|---|---|
| A. | "Veegum" T | 1.50 |
| | Water | 24.00 |
| B. | Ammonium hydroxide | 6.00 |
| | 1,1,1-trichloroethane | 18.00 |
| | Ethanol | 7.50 |
| | "Tergitol" NPX | 18.00 |
| | Propellent 12/114—70/30 | 25.00 |

1. Add the "Veegum" to the water slowly, agitating continually until smooth.
2. Add B to A and mix until uniform.
3. Package in aerosol dispensers.

Shake before using (general instruction). Oven should be cool and empty. Be careful not to spray on painted surfaces or rubber or plastic tile (trichloroethane may attack paint and tile). Contains alcohol. Do not use near flame or on hot surface (general precautionary instruction). Hold can upright and point valve away from eyes, holding can about a foot away from the oven surface. Spray oven surface, leaving a layer of foam. Allow foam to stay and work for 15 minutes. This allows time for solvent and surfactant to attack grease. Wipe oven clean with coarse damp cloth or sponge. Reapply to stubborn spots if needed.

**Rust Remover**

| | |
|---|---|
| Phosphoric acid (85%) | 47.2 |

| | |
|---|---|
| Butyl "Cellosolve" | 16.0 |
| "Triton" X-100 | 2.0 |
| Water | 34.8 |

Add in order phosphoric acid, butyl "Cellosolve", and "Triton" X-100 to the water. If cold water is used, preblend the "Triton" X-100 with 3 parts warm water before adding. Dilute 3 to 1 for use.

---

**Corrosion Preventive**

| | |
|---|---|
| *Arachidyl-behenyl primary amine | 6.00 |
| *Tetrapropenylsuccinic anhydride | 4.58 |
| *Water | 0.46 |
| Lubricating oil, grade 1010 | 11.96 |
| Oxidation inhibitor | 0.15 |
| n-Heptane | 53.32 |
| *Ethanol | 23.53 |

*React at about 70°C.

---

**Oil Rust Prevention**

| | |
|---|---|
| Preservative engine oils | Add "Atpet" 20 2-4% |
| Gear lubricants | " " " 2-4% |
| Slushing compounds | " " " 2-4% |
| Preservative oils for machinery and tools | " " " 2-4% |
| Greases | " " " 3-5% |

---

**Photographing on Anodized Aluminum**

Black

| | |
|---|---|
| Distilled water (38-65°C, 100-150°F) | 850 |
| Citric acid | 10 |
| Thiourea | 5 |
| Aluminum sulfate | 22 |
| Sodium chloride | 10 |
| Phloroglucinol (photo grade) | 30 |
| Butyl "Cellosolve" | 150 |
| Cool to 38°C or less and dissolve: | |
| p-diazo-diethylaniline 0.5 zinc chloride | 40 |

Red

| | |
|---|---|
| Distilled water (38-65°C, 100-150°F) | 850 |
| Boric acid | 31 |

| | |
|---|---|
| Citric acid | 90 |
| Thiourea | 46 |
| Aluminum sulfate | 9 |
| Chloro-aceto-pyrocatechol | 10 |
| Butyl "Cellosolve" | 150 |
| Cool to 38°C or less and add: | |
| p-diazo-diethylaniline 0.5 zinc chloride | 24 |

Yellow

| | |
|---|---|
| Distilled water (38-65°C, 100-150°F) | 850 |
| Citric acid | 10 |
| Thiourea | 5 |
| Aluminum sulfate | 22 |
| Sodium chloride | 10 |
| Acetoacetanilid | 15 |
| Butyl "Cellosolve" | 150 |
| Cool to 38°C or less and add: | |
| p-diazo-diethylaniline 0.5 zinc chloride | 40 |

Blue

| | |
|---|---|
| Distilled water (38-65°C, 100-150°F) | 850 |
| Citric acid | 10 |
| Thiourea | 5 |
| Aluminum sulfate | 22 |
| Sodium chloride | 10 |
| 2, 3-Dihydroxynaphthalene | 30 |
| Butyl "Cellosolve" | 150 |
| Cool to 38°C or less and add: | |
| p-diazo-diethylaniline 0.5 zinc chloride | 30 |

Note: Light blue may be made by decreasing amount of diazo by 10 g/l.
Dark blue may be made by increasing diazo by 10 g/l.

Solution should be stored in a cool dark place. Use within 48 hr after adding the light-sensitive (diazo) ingredient. Solutions may be mixed together to make secondary colors. Hundreds of colors are also possible by using the large varieties of diazos and couplers available commerically.

Anodizing Process

1. Chemically clean the aluminum panel and rinse in deionized water.
2. Anodize in 18% sulfuric acid at 24-27°C for at least 30 min at a current density of 1.3 amp/$dm^2$ (12 amp/$ft^2$) making the workpiece the anode.
3. Rinse in deionized water and dry in clean warm air.

---

**Sensitizing Anodized Aluminum**

Excellent results have been achieved by flowing the solution on freshly anodized and dried aluminum surfaces which may then be force dried using a fan or by placing in an oven at 38 to 54°C. The dried film has a yellow appearance. A darkroom is not necessary for sensitizing, but subdued (yellow) light should be used to prevent premature exposure.

---

**Exposing the Sensitized Aluminum**

The emulsion side of a photopositive transparency is placed in intimate contact with the sensitized face of the aluminum. Ultraviolet light in the form of an arc, a UV lamp or sunlight may be used for exposure. Length of time for exposure is usually determined experimentally and varies with the amount and type of diazo compounds used in the formulation. A good indication that sufficient exposure has been achieved is the apparent bleaching of the yellow diazo color in the exposed areas.

---

**Developing the Color**

One or two seconds' exposure to a moist ammonia atmosphere will cause the formation of a dense insoluble dye in the unexposed areas of the aluminum. The background sometimes contains a white precipitate. Dyes generally change color with *p*H. For this reason the final color appears only after the ammonia has completely evaporated.

---

**Sealing the Pores**

Conventional sealing using either nickel acetate solution for 5 min or boiling water for 30 min will dissolve the undesirable byproducts in the film and permanently seal the image. Light fastness equal to that achieved with organic dyeing can be expected.

## Nickel Plating

| | |
|---|---|
| Nickel sulfamate | 375 g/l |
| Nickel chloride | 10 g/l |
| Total nickel | 81 g/l |
| Boric acid—$H_3BO_4$ | 30 g/l |
| Temperature | 46-48°C |
| *p*H (electrometric) | 3.8-4.2 |
| Cathode current density | 2.15 amp/dm² (20 amp/ft²) |
| Agitation | mild mechanical stirring |

## Bright Nickel Plating Solutions

*Solution composition, g/l*

| *No.* | *$NiSO_4 \cdot 7H_2O$* | *$NiCl_2 \cdot 6H_2O$* | *$H_3BO_3$* | *HCK** | *Chloramine* | *Quinoline* |
|---|---|---|---|---|---|---|
| 1 | 200 | 15 | 30 | — | 1 | 0.025 |
| 2 | 200 | 15 | 30 | 1 | — | 0.025 |

*Plating Conditions*

| *No.* | *Cathode current density, amp/dm²* | *Temp., °C* | *pH* | *Current effic. %* | *Brightness (scale units)* |
|---|---|---|---|---|---|
| 1 | 1-6.5 | 40 | 2-5 | 95-99 | 9-10 |
| 2 | 1.5-4 | 40 | 3-5 | 95-99 | 9-10 |

*Mixture of sodium salts of 2,6- and 2,7-naphthalene disulfonic acids.

## Plating With Silver Copper

| | |
|---|---|
| Silver | 1.29 g/l |
| Copper | 47.70 " |
| Free KCN | 1.70 " |
| Pot. carbonate | 18.20 " |
| pH | 12.40 |

Plate at 1-3½ amp./sq. ft.

## White Gold Plating

No. 1

| | |
|---|---|
| Gold, as gold potassium cyanide | 0.125 oz/gal |
| Nickel, as nickel potassium cyanide | 0.675 " |
| Free cyanide | 0.050 " |

150-160°F, 30-200 amp/ft², Stainless steel anodes.

No. 2

| | |
|---|---|
| Gold, as gold potassium cyanide | 0.650 oz/gal |
| Potassium stannate | 11.500 " |
| Potassium hydroxide | 3.000 " |
| Potassium cyanide | 2.000 " |

150-160°F, 30-50 amp/ft$^2$, Stainless steel anodes.

## Rhodium Plating

| | For 0.0001-in. Plate | For 0.0003 in.+Plate |
|---|---|---|
| Rhodium sulphate TP | 20 g | 40 g |
| Sulphuric acid, C.P. | 100 ml | 100 ml |
| Distilled water | to make 1 gal | to make 1 gal |
| Current density | 10-50 ASF | 10-40 ASF |
| Replenishment, 5 g | per 5 amp.-hr | per 4.5 amp.-hr |

Tank: Glass, plastic-lined steel ("Koroseal"), rigid PVC such as "Boltaron" and "Lucite".

Anodes: Platinum or Platanodes (0.04-in. platinum-plated titanium sheet). Anode to cathode ratio, 1/2 to 1 minimum, preferably 1 to 1.

Leach all plastic tanks and especially cemented lucite joints, with warm (130°F) 10% by volume sulfuric acid for 24 hr. Rinse well. Fill the clean tank half full with distilled or deionized water. Add the required amount of reagent-grade sulfuric acid slowly, with agitation. Mix well. Add the rhodium sulfate TP with stirring. Mix well. Add distilled water to bring liquid level to operating volume. Mix well.

Operating Data

| | |
|---|---|
| Temperature | 110-120°F |
| Agitation | Preferred for plates over 0.000020 in. thick. |
| Concentration | Dependent on thickness required. |
| Current Density | 10-30 ASF |

## Gold and Silver Recovery from Stripping Solutions

Gold and silver can be recovered from cyanide stripping solutions by plating out on stainless steel cathodes, or by precipitating with zinc dust. Gold does not dissolve in the concentrated sulfuric acid strip but precipitates, so that recovery is simply a matter of removing the mud. Silver can be precipitated from the sulfuric-nitric acid strip by diluting the strip with water and then adding a sodium chloride solution to precipitate silver chloride.

Chapter XIV

# PAPER

## Paper Coatings

| | No. 1 | No. 2 | No. 3 | No. 4 | No. 5 | No. 6 |
|---|---|---|---|---|---|---|
| Ethylcellulose | | | | | | |
| (10 cps) | 15 | 10 | 20 | 20 | 30 | 27.0 |
| "Castorwax" | 20 | 44 | 28 | 14 | 24 | 38.5 |
| Paraffin | 25 | 28 | 32 | 26 | 14 | 11.5 |
| Resin | | | | | | |
| ("Super-Beckacite" 2000) | 15 | 18 | — | — | — | — |
| ("Staybelite") | 20 | — | — | — | — | — |
| ("Dow 276-V9") | — | — | 20 | 40 | — | — |
| (Polypale ester No. 21 or No. 10) | — | — | — | — | 28 | — |
| Plasticizer | | | | | | |
| (Diamyl phthalate) | 5 | — | — | — | — | — |

Uses

No. 1. Coated and heat-sealing food packaging; soap wrappers; shelf paper; label coating.

No. 2. Bread-wrapper coating.

No. 3. Food packaging with very low moisture and vapor transmission.

No. 5. Coating with unusual grease proofness.

No. 6. Balanced formulation to give clear, hard, flexible and tough coating with good adhesion to paper. Designed to meet T454-m44 of TAPPI standards.

| | No. 7 | No. 5 | No. 9 |
|---|---|---|---|
| n-Butyl methacrylate | 37.5 | 34.0 | 40.0 |
| "Castorwax" | 37.5 | 34.0 | 20.0 |
| Paraffin (135°F) | 25.0 | 23.0 | 20.0 |
| "Santicizer" B-16 | — | 9.0 | — |
| Rosin | — | — | 20.0 |

**Hot Melt Paper Strippable Coating**

| | |
|---|---|
| "Ethocel"—50 cps | 25 |
| Mineral oil | 50-60 |
| "Castorwax" | 1-3 |
| "Baker's" 15 oil | 9-13 |
| "Estynox" 140 | 1-9 |
| Antioxidant | 0.25-2.0 |
| Color stabilizer | 2.0 |

CHAPTER XV

# POLISH

### Asphalt-Tile Wax Polish, Antislip

| | |
|---|---|
| "Efton" UC | 12.0 |
| "Duron" 195-D | 9.5 |
| "Duron" 180-D | 2.5 |
| Stearic acid, double pressed | 8.0 |
| "Amsco" 46 | 15.0 |
| Triethanolamine | 3.0 |
| Soda ash | 1.0 |
| Water | 49.0 |

Melt the "Efton" UC, the "Durons" and Stearic acid together in one container at a temperature of 230 to 250°F. In a separate container, heat the water in which the triethanolamine and soda ash have been dissolved and adjust the temperature to 205-210°F. Adjust the temperature of the wax melt to about 210°F by shutting off the heat while continuing ot agitate. Then add the "Amsco" 46 to the wax melt and agitate until uniform and clear. To maintain solution of the waxes in the solvent, it may be necessary to reheat slightly. Next, start effective paddle or turbine-type agitation in the water kettle and slowly add the wax blend. Heat and agitate until a smooth, creamy emulsion is obtained. Cool with agitation to about 195°F and pour into suitable containers. Allow the filled containers to stand undisturbed for further cooling and paste formation.

By increasing the proportion of water in this formula, a softer, creamier, yet stable product can be obtained.

---

### Nonbuffable Wood Floor Polish

Solvent Type

| | |
|---|---|
| "Acryloid" B-67 (100%) | 7.5 |
| "Durez" 225 | 1.5 |
| "Hoechst" Wax "V" | 0.5 |
| "Halowax" 1013 | 0.5 |
| "Amsco" 46 | 90.0 |

**Rubless Floor Polish**

No. 1

| | |
|---|---|
| "Ceramer" 67 | 75.0 |
| "Epolene" E-12 | 20.0 |
| "Starwax" 100 | 5.0 |
| Oleic acid | 2.0 |
| Zinc octoate (18% Zinc) | 1.5 |
| Diethylethanolamine | 2.5 |
| Ammonium hydroxide (26° Be) | 6.0 |
| "KP"-140 | 0.5 by volume |
| "FC"-128 | 1.0 by volume |
| Water (hot) | 224.0 |
| Water (cold) | 398.0 |

This formula produces a polish of 14% solids.

Melt the solid components, including oleic acid, and bring to a temperature of 235 to 240°F. Add the zinc octoate and DEEA. Cook for 5 min. In the meantime heat 224 parts of water to 205°F. This produces a 30% solids emulsion. When the wax melt is ready, add the $NH_4OH$ to the hot water and then introduce the wax melt with rapid agitation.

After the 30% concentrate has been prepared. it should be chilled rapidly by adding it, with rapid stirring, to 398 parts of cold water to yield an emulsion of 14% solids. At no time should the temperature of the concentrate be allowed to drift below 150°F and the temperature of the diluted emulsion should be below 130°F after all the concentrate has been added.

When the final emulsion has cooled to room temperature, add the FC-128 and then carefully stir in the "KP-140". For good levelling the "KP-140" must be completely dissolved. This formula produces a polish which can be recoated within an hour after the first coat has dried. It has excellent initial gloss and buffability. The use of a small amount of oleic acid aids in plasticizing the film and produces a polish with an excellent sheen and good wearing properties. Scuffing is normal for a wax polish and scuff marks are easily buffed out. The film shows extremely good black heel mark resistance and good detergent resistance after aging.

No. 2

Wax Emulsion

| | |
|---|---|
| "Petrolite" C-7500 | 50 |

| | |
|---|---|
| "Petrolite" C-700 | 15 |
| "AC"-680 | 10 |
| "AC"-540 | 25 |
| Oleic acid | 6 |
| KOH (85%) | 2 |
| Morpholine | 10 |
| Water (hot) | 247 |
| Water (cold) | to desired final solids |

Levelling Resin Solution

| | |
|---|---|
| Levelling resin | 90 |
| Tri-Butoxyethyl phosphate | 10 |
| Ammoniated water | to same solids as wax emulsion |

Polish

| | |
|---|---|
| Wax emulsion | 80 |
| Levelling resin solution | 20 |
| FC-128 (1% solution) | 1 |

*Wax Emulsion*: The emulsion is produced by melting the waxes, polyethylene and oleic acid together. This mixture is brought to a temperature of 240 to 250°F and the KOH is then added slowly as a hot saturated solution. It is recommended that distilled water or demineralized water containing no carbonates be used to dissolve the KOH. The resulting mixture is cooked for at least 15 min. During this period the temperature can be allowed to drop back to 225 to 230°F. At the end of this time most of the foam has subsided and a clear, slightly viscous mixture results. The morpholine is added and cooked at 225 to 230°F for an additional 5 min. This mixture is then raised to a temperature of 230°F and poured into 247 parts of water at 205°F with rapid agitation. This produces an emulsion at 30% solids concentration. Do not let this concentrate cool below 180°F at any time. For maximum initial gloss the concentrate should be introduced into the cold dilution water with rapid agitation. The temperature differential between the hot concentrated emulsion and the cold dilution water should be such that the temperature of the final diluted emulsion is not greater than 130°F. Obviously this temperature differential is a function of the volume of water required for the final dilution. If the concentrate is to be reduced to 16% solids or less room temperature water is cold enough. If the final solids is to be higher than 16% it may be necessary to cool the dilution water somewhat below room temperature.

*Levelling Resin Solution*: The levelling resin solution is prepared as follows: Heat the water to 190 to 195°F, add half the required amount of ammonia to the hot water. Use 28° Be'$NH_4OH$. Because of the volatility of ammonia at these elevated temperatures, it is recommended that a covered vessel be used. The amount of ammonia required will vary with the acid number of the resin to be dissolved and this information is available from the resin manufacturer. Usually about 20 lb of ammonia is required for 100 lb of resin.

After the ammonia has been well mixed with water, the resin is added and agitation continued. When the resin is well dispersed, add balance of ammonia. When all the resin is dissolved, adjust pH of finished solution to 8 to 9.

The finished solution can be stored indefinitely without deterioration, provided steps are taken to avoid the loss of ammonia.

This is a high quality, relatively low-cost polish. Initial gloss is unusually good for a wax-based polish, and buffing characteristics are excellent. Long-term floor tests show high resistance to black heel marking and an exceptionally low degree of scuffing. Powdering is no problem with this polish. Slip resistance is high and wearing properties are outstanding. The film is practically colorless.

No. 3

| Emulsion | |
|---|---|
| "Petrolite" C-8500 | 50 |
| "Epolene" E-10 | 30 |
| "Ceramer" 67 | 20 |
| Oleic acid | 3 |
| KOH (85%) | 1 |
| Diethylethanolamine | 7 |
| Water (205-208°F) | 233 |
| Water (Cold) | 395 |
| **Levelling Resin Solution** | |
| "Durez" 19788 | 90 |
| Tributoxethylphosphate | 10 |
| $NH_4OH$ (26° Be) | 25 |
| Water | 590 |
| **Polish** | |
| Emulsion | 80 |
| Levelling resin solution | 20 |

Heat waxes and oleic acid to 235 to 240°F. Add saturated solution KOH slowly and react for 15 min. Add diethylethanolamine and react for 5 to 10 min. This wax mixture is now poured into the 233 parts of hot water using good agitation. This results in an emulsion at 30% solids. Care should be taken not to allow this concentrated emulon to cool below 150°F. The hot concentrated emulsion is now introduced into the 395 parts of cold water with good agitation. The temperature of the finished emulsion should be 130°F or less.

This is a high-quality polish.

It has high initial gloss and excellent buffing characteristics. Among its more outstanding characteristics are its high resistance to wear, particularly as regards black heel marking, its lack of slipperiness, and its complete resistance to water damage after 24 hr aging. It is not recommended that this polish be recoated until the initial coat has aged for 24 to 48 hr. During this period there is a possibility of disturbing the first coat by applying a second coat.

This polish works well on all types of composition flooring but is particularly effective on asphalt, vinyl asbestos, and rubber.

No. 4

| | |
|---|---|
| "Petrolite" C-9500 | 1.1 |
| "Petrolite" C-700 | 3.3 |
| Paraffin wax (140/142 AMP) | 5.5 |
| Calcium stearate | 1.1 |
| Stoddard solvent | 89.0 |

The solvent is divided into two equal parts. One part is heated to 210 to 220°F with the waxes, and the mixture is stirred and held at that temperature for 10 min, or until complete solution is obtained. The second part of the solvent is then added cold, with stirring. The mixture is now cooled, with constant stirring, to 90°F. At this temperature it is poured into containers.

**Caution:** The recommended operating temperatures are above the flash point of the solvent. Avoid open flames or sparks which could cause ignition of the solvent.

It is important in preparing this polish to maintain the solvent wax mixture at 210 to 220°F for a sufficient length of time to get all the calcium stearate into solution. The calcium stearate acts as a crystal-growth inhibitor producing finer crystals which aids in binding the solvent. Care must be used in employing calcium stearate, since too great a quantity will cause complete gelling. This polish performs well on the floor. It spreads easily and buffs to a good shine. It shows excellent resistance to traffic and is adequately slip-resistant.

No. 5

Part A

| | |
|---|---|
| "Petrolite" C-7500 wax | 100.0 |
| Oleic acid | 12.0 |
| KOH (solid, technical grade) | 2.0 |
| Morpholine | 10.0 |
| Water | 835.0 |

Part B

| | |
|---|---|
| "Dow Corning" DC-200 silicone, 350 ctks | 4.0 |
| Stoddard solvent | 19.0 |
| Kerosene | 2.0 |
| Oleic acid | 2.5 |

Part C

| | |
|---|---|
| Morpholine | 1.5 |
| Water | 16.0 |

Part D

| | |
|---|---|
| "Snow Floss" | 14.0 |
| Water | 31.0 |

To prepare Part A, melt the wax and oleic and bring temperature up to 230 to 240°F. Dissolve the KOH in the minimum amount of distilled or deionized water required for complete solution. Heat to boiling and add slowly to the hot wax with good agitation. After all the KOH solution has been added, cook the resulting mixture for 15 min at 225 to 230°F. Then, add the morpholine and allow 5 min reaction time. This mixture, at 230°F, is then introduced into the water, which is held at 200 to 205°F, with high-speed stirring. The resulting emulsion should be cooled to room temperature as rapidly as possible. To prepare the finished polish, mix Part B, at room temperature, using a high speed stirrer. Slowly add Part C. Continue mixing while adding Part D and finally 10 parts of Part A.

This formulation produces an excellent liquid polish for use on both old and new cars. There is enough abrasive ("Snow Floss") present to give good cleaning action, and the wax and silicone impart a high polish which is long lasting. The film will retain its water repellancy for a considerable period. In addition to the cleaning action imparted by the abrasive, the solvent in the formula will remove any road oil or grease on the car and the soaps will remove any water soluble soil left on the surface. It is recommended that the car be washed and dried before applying the polish. Also, the polish should not be applied if the car has been standing for any length of time in the hot sun.

No. 6

The following formula produces bright, non-slip polish films. The leveling is excellent. The polish has particularly good resistance against dirt pick-up and black heel marking. Gloss and color-brightening retention is excellent. Low-cost dry maintenance by brush-buffing, with occasional damp mopping is sufficient to extend the service life of the polish over many weeks in heavy, even wet, traffic.

| | |
|---|---|
| "Durmont"/"Duron" mother emulsion 6550* | 80.00 |
| Ammoniacal solution of "Shanco" 334 resin | 20.00 |
| 2-Pyrrolidon | 0.25 |
| "KP"-140 | 0.25 |

Melt the "Durmont", "Duron", polyethylene and oleic acid together, and cook with the KOH at about 240°F for 20-30 min, until the batch becomes clear and foaming has ceased. Then add the morpholine. Stir until clear. Adjust the melt temperature to 240 to 245°F. Pour the melt into water of 203 to 205°F with strong agitation. Avoid foaming and surface boiling, as these may cause graininess and to loss foaming and surface boiling, as these may cause graininess and loss of film clarity and gloss. Continue heat and agitation for 1 to 2 min after all the wax has been dispersed. Then turn on the cooling water and agitate slowly until the batch reaches room temperature.

---

*"Durmont"/"Duron" mother emulsion 6550

| | |
|---|---|
| "Durmont" E | 27.0 |
| "Duron" 195-D | 33.0 |
| "AC"-680 emulsifiable polyethylene | 40.0 |
| Oleic acid | 8.5 |
| KOH (40% aqueous solution) | 3.0 |
| Morpholine | 11.0 |
| Water | 634.5 |

No. 7

This formula yields a product of exceptionally white, translucent appearance which is nondiscoloring because it makes use of a water-clear leveling resin and an emulsifier system which, in contrast to morpholine oleate, does not stain certain sensitive white materials used as fillers in flexible tile. Its initial gloss and leveling are both rated as "excellent". Although 100% synthetic, the product rates with any wax on dry maintainability and resistance to heel marking.

| | |
|---|---|
| Water at room temperature | 40.3 |
| "Durmont"/"Efton"/polyethylene base No. 6712* at 30% N.V. | 38.7 |
| Ammoniacal solution of "SMA"-2625A resin | 20.0 |
| "KP"-140 | 0.5 |
| "Carbitol" | 1.0 |
| FC-128, 1% solution | 1.0 |

Add with light stirring in the order as shown. To incorporate the plasticizer system, use high speed agitation for 30 min.

Melt "Efton", "Durmont", "Epolene", oleic acid and "Tween" 80 together, raise the temperature to 240°F and react *without stirring* with the KOH solution for 20 to 30 min until the batch clears and foaming ceases. Then add the mixture of 3-methoxy propylamine and amino-methyl-propanol. Stir until clear. Pour the melt at 240 to 245°F into water at 203 to 205°F with strong agitation. Avoid foaming or surface boiling. Temperatures are important! Continue heat and agitation about 2 min after all the wax has been dispersed. Then turn on the cooling water and agitate slowly to room temperature. Cover the batch to avoid skin formation.

---

*"Durmont"/"Efton"/polyethylene base No. 6712

| | |
|---|---|
| "Efton" D super | 50.0 |
| "Durmont" E | 10.0 |
| "Epolene" E-10 | 40.0 |
| Oleic acid | 4.0 |
| "Tween" 80 | 1.0 |
| KOH, 45% aqueous solution | 4.5 |
| Amino-methyl-propanol | 0.5 |
| 3-Methoxy propylamine | 5.0 |
| Water | 243.3 |

---

**Floor Polish Leveling Solution**

| | |
|---|---|
| "KP"-140 | 5.0 |
| Dibutyl phthalate | 3.5 |
| "Carbitol" | 5.0 |
| Methyl pyrrolidone | 5.0 |
| "Renex" No. 690 | 5.0 |
| Water | 5.0 |
| 1% aqueous solution of "FC"-1% aqueus solution of "FC"-128 | 3.0 |

---

**Detergent-Resistant Floor Finish**

| | |
|---|---|
| "Rhoplex" B336 | 67.5 |
| Wax-polyethylene base 6866* | 25.0 |
| Ammoniacal solution of "Durez" 19788 | 7.5 |
| "KP"-140 or equiv. TBEP | 1.0 |
| "Carbitol" | 1.0 |
| FC-128 at 1% | 0.5 |

---

*Wax-Polyethylene base

| | |
|---|---|
| "Duron" 180-D | 30.0 |
| "Durmont" E | 30.0 |
| "Epolene" E-15 | 40.0 |
| "Acintol" FA3 | 5.0 |
| KOH at 43% | 3.0 |
| 3-Methoxy propylamine | 15.0 |
| Ammonia 28% | 226.0 |
| Water at 195°F | 402.0 |
| Water at room temperature | |

Melt the waxes, "Epolene" and fatty acid together, add KOH solution at 225°F, and react without stirring 20 to 30 min until batch clears and foaming ceases. Add the 3-MPA. Readjust melt to 255°F. Bring water temperature (226 g portion) to maximum 200°F. Start high-speed agitation and add the $NH_3OH$ to the water. Slowly pour melt into the ammoniated water with consistent agitation. Hold at heat for 1 to 2 min, then add balance of water. Cool to room temperature and store or use in polymer blend.

**Scuff-Resistant Floor Polish**

| | |
|---|---|
| "RWL"-112 latex (40%) | 198.0 |
| "RWL"-201 latex (40%) | 50.0 |
| Water | 412.0 |
| "Shanco" 1165-S resin solution (15%) | 180.0 |
| "A-C" polyethylene 629 emulsion (15%) | 160.0 |
| Tributoxyethyl phosphate | 6.0 |
| "Igepal" CO-630 | 3.0 |
| Diethylene glycol | 4.0 |
| Formalin | 1.5 |

Add the water to the polishing mixing kettle. Then add the "RWL"-201 latex with stirring. Adjust pH to 8.0 to 9.0 with 28% ammonium hydroxide. Then add the "RWL"-112 latex, the resin solution, and the polyethylene emulsion with stirring. Combine the plasticizers and add slowly with adequate agitation to the latex-resin-wax mixture. Stir for 1/2 hr. Then add the formalin and adjust the final pH to 8.8 to 9.0 with 28% ammonium hydroxide.

**Semibuffable Floor Finish**

| | |
|---|---|
| Wax base No. 7165* | 25.0 |
| "UBS" polymer WC-30 at 14% solids | 62.5 |
| Ammoniacal solution of "Durez" 19788 | 12.5 |
| Methyl "Carbitol" | 2.0 |
| "KP"-140 | 1.0 |
| 1% Aqueous solution of FC-128 | .1.0 |

*Wax—Polyethylene Emulsion Base No. 7165

| | |
|---|---|
| "RUHRWAX" AV–1551 | 25.0 g |
| "DURON" 180D | 10.0 g |
| "DURON" 195D | 35.0 g |
| "AC"–680 | 30.0 g |
| Oleic acid | 4.0 g |
| KOH at 45% | 3.0 g |
| Morpholine | 11.0 g |
| Water at 205°F | 232.0 g |
| Water at room temperature | 400.0 g |

Melt the waxes, the "AC"-680 and the oleic acid together, add KOH solution at approx. 230°F and react for 20 to 30 min with mild agitation, until the batch clears and foaming ceases. Increase the temperature to 250°F and add the Morpholine. Stir until clear. Maintaining the wax melt at 250 to 255°F, pour it slowly with rapid agitation into the 232 g of water at 205°F. Continue agitation and heating for an additional 1 to 2 min after all the wax has been poured. Cool rapidly to room temperature with mild agitation, and then add the emulsion to the rest of the water at room temperature.

---

**Nonbuffable, Detergent-Resistant Floor Polish**

| | | |
|---|---|---|
| "Rhoplex" B-336 (15% solids) | | 80.0 |
| Alkali-soluble resin "Shanco" 334 (15% solids in ammonia) | | 5.0 |
| "Epolene" E-10 (15% solids in aqueous morpholine oleate) | | 15.0 |
| "Carbitol" | | 1.5 |
| Ethylene glycol | | 1.5 |
| Tributoxyethyl phosphate | premix | 0.5 |
| "FC"-128 (1% solution) | | 0.4 |

Add the ingredients in the order listed above. Predilute a mixture of the last four ingredients with at least an equal amount of water (the water to be calculated as part of that used for dilution of the other components). Incorporate the resulting solution slowly, with good agitation. The addition of 500 ppm of formaldehyde is recommended as preservative.

---

**Paste Floor Polish and Cleaner**

| | | |
|---|---|---|
| Part 1. | "Duroxon" J-324 | 16.0 |
| | "Efton" UC | 15.0 |
| | "Mycrox" 114 | 4.0 |
| | "AC" polyethylene 629 | 5.0 |
| | Stearic acid, double-pressed | 8.0 |
| | Oleic acid, double-distilled | 5.0 |
| | "Amsco" Mintrol spirits | 25.0 |
| Part 2. | Amino-methyl-propanol | 6.0 g |
| | Water | 115.0 g |
| | Caustic potash 45% | 2.0 g |
| | Borax, 10 mol | 0.5 g |

All components of Part 1, except the mintrol spirits, are heated together and melted at a temperature of approximately 225°F. Then the mintrol spirit is slowly added and the temperature adjusted to about

200°F. Then Part 2 at 200°F is poured with agitation into Part 1. Continue stirring until temperature is 185°F, then pour into containers.

Directions for Single-Step Floor Maintenance Program

1. Use the regular rotating brush power polishing equipment, equipped with wire brush. Stuff the hole in the center of the wire brush with wadding or steel wool to form a level surface.
2. Place a No. 3 steel-wool pad on the brush surface and spread a quantity of the paste gel ($\frac{1}{4}$ to $\frac{1}{2}$ lb) on the center of the pad.
3. Use the machine with this steel-wool assembly in the same manner as for polishing the floor. This cleans the floor and deposits a protective wax-acrylic finish at the same time.
4. When additional paste gel compound is required, tilt the machine back and place the compound directly underneath the center of the pad. Both sides of the steel wool pad can be used. Reverse the pad when one side becomes clogged with soil.
5. To finish up the operation, buff the entire floor area to a high gleam using clean No. 3 steel-wool pad.

If floor has had an excessive buildup of old wax, it should be stripped thoroughly by conventional methods to prepare it for the subsequent maintenance program by the above singe-step operation.

---

**Floor Polish Stripper Concentrates**

No. 1

| | |
|---|---|
| Ethylene glycol monobutyl ether | 2 |
| Phosphoric acid (85%) | 1 |
| "Igepal" CO-730 | 1 |
| Phenyl glycol ethers | 4 |
| Phosphoric acid (85%) | 2 |
| "Igepal" CO-730 | 1 |

Mix the solvent with the "Igepal" CO-730. Add the phosphoric acid with stirring.

These formulations, diluted (1:40 to 1:50) with hot water (120 to 140°F), will readily remove the alkali-resistant, acid-sensitive polishes. Use in a manner similar to that employed with ordinary alkaline-based wax strippers. For example, apply the hot solution of diluted stripper to the floor. With brushes or stripping machines, work the stripping solution into the polish film. The film readily disintegrates. Mop up the residue without allowing the floor to dry. Rinse the floor thoroughly with warm water and mop dry. One gallon of dilute stripper will remove approximately 500 sq. of old polish.

No. 2

| | |
|---|---|
| "Triton" QS-44 | 4.0 |

| | |
|---|---|
| Sodium hydroxide | 0.8 |
| Tetrapotassium pyrophosphate | 12.0 |
| Tetrasodium pyrophosphate | 8.0 |
| Sodium metasilicate, anhydrous | 0.4 |
| Water | 74.8 |

---

**Scratched-Wood Polish**

| | | |
|---|---|---|
| A. | "Veegum" T | 1.50 |
| | Water | 25.50 |
| B. | Stearic acid | 4.50 |
| | Carnauba wax | 16.50 |
| | Beeswax, yellow | 16.50 |
| C. | Triethanolamine | 2.75 |
| D. | Mineral spirits | 27.00 |
| E. | Color | 0.50 |
| | Water | 5.25 |

1. Add the "Veegum" T to the water slowly, agitating continually until smooth.
2. Heat B to 90°C, add C and maintain temperature.
3. Add D, maintaining 90° temperature, avoiding open flames.
4. Add 1 to 3, mixing until uniform.
5. Disperse E in hot water and add to 4, continuing to mix until uniform by processing through a roller mill.

Apply to damaged areas with a soft cloth and rub into area. Buff with a clean dry cloth.

---

**Furniture and Auto Polish**

| | |
|---|---|
| Light mineral oil | 48.0 |
| "Surfactol" 365 | 8.0 |
| Oleic acid | 6.6 |
| Monoethanolamine | 0.5 |
| Water | 60.0 |

---

**Aerosol Emulsion Furniture Cream Polish**

Concentrate

| | |
|---|---|
| "Duroxon" H-111 | 2.5 |
| "Duroxon" E-321 | 1.5 |
| "Emulphor" ON-870 | 2.0 |
| AMP | 0.5 |
| Wood turpentine | 25.0 |
| Water | 68.5 |

Aerosol Fill—8-oz Lacquer-lined Can

| | |
|---|---|
| Concentrate as above | 125 ml |
| Propellant No. 11/12, mix 50/50 | 50 ml |

---

**Aerosol Furniture Cleaner-Polish**

| | |
|---|---|
| (1) "Union Carbide" silicone (L-45 350 CSTK) | 3.0 |
| (2) "Crown" wax 23 | 2.0 |
| (3) Oleic acid | 2.0 |

| | |
|---|---|
| (4) Triethanolamine | 1.0 |
| (5) Mineral spirits | 20.0 |
| (6) Water | 59.3 |
| (7) "Maprofix" TLS-500 | 2.3 |
| (8) "Onyxol" WW | 0.4 |
| (9) "Carbopol" 934 (2% sol.) | 10.0 |

Heat (1), (2), and (3) in approx. 20% of (5) until clear, then add rest of (5) and hold at 158°F until ready for use. Add (4) to (6). Hot solvent solution. Slowly to water with vigorous agitation, then add (9), (7), and (8).

Package in foam type aerosol container. Using 18.5 parts "Freon" 12 per 100 parts of foam.

---

## High-Gloss Furniture Cream Polish

The formula described in the following yields a stable furniture cream in liquid form for packaging in plastic squeeze bottles or glass bottles. The latter may be equipped with a plunger-type dispenser. This polish has demonstrated outstanding gloss and gloss retention, with a minimum of wicking. It may be applied to wood, plastic, "Formica" or metal surfaces with equally satisfactory results. The rub out is effortless.

| | | |
|---|---|---|
| 1. | "Duroxon" B-120 | 5.00 |
| | "Arlacel" C (Atlas) | 1.50 |
| | Silicone fluid SF-96/1000 ctks. | 4.00 |
| | "Amsco" mineral spirits | 20.00 |
| 2. | Water at 190°F | 68.25 |
| | "Carbopol" 934 | 0.25 |
| 3. | Sodium hydroxide, 10% aqueous solution | 1.00 |

Dissolve the "Duroxon", "Arlacel" C and silicone fluid in the mineral spirits at 190 to 200°F. Disperse the "Carbopol" in the water in a separate container at 190 to 200°F with high-shear agitation. Pour 1 into 2 with agitation. Add 3 and continue high-speed agitation to maximum 100°F or room temperature.

---

## Emulsion Cleaner Polish—(W/O Type)

| | |
|---|---|
| "Dow Corning" 531 fluid | 4.50 |
| "Dow Corning" 530 fluid | 0.75 |
| "Span" 80 | 1.00 |
| "Kaopolite" SF-O | 10.00 |
| Stoddard solvent | 15.75 |
| Kerosene | 10.00 |
| Water | 58.00 |

Add all ingredients except water and mix until uniform. Add water slowly and mix at high speed until a suitable emulsion is produced. Package.

The above formula produces a water-in-oil, creamy, thixotropic emulsion. On standing, the emulsion becomes thinner. If isopropyl alcohol or polar solvents are added to the above formula, the emulsion consistency is reduced to a thin liquid. An excess of polar solvent may cause the abrasive to settle.

### Spray and Wipe Aerosol Polishes

| | No. 1 | No. 2 | No. 3 | No. 4 |
|---|---|---|---|---|
| Union Carbide LE-461 | 10.0 | 7.0 | 12.5 | 12.5 |
| Union Carbide LE-462 | | | 2.5 | 2.5 |
| Triethanolamine lauryl sulfate | | | | 1.5 |
| Concord "Co-Wax" | | 3.3 | 6.0 | 6.0 |
| Distilled water | 90.0 | 89.7 | 79.0 | 77.5 |

To prepare a 15% solids emulsion of "Co-Wax" and water, heat the water to 180-210°F, melt "Co-Wax", and slowly pour the molten wax into the heated water with vigorous agitation. Cool the emulsion to room temperature. Stir in Union Carbide LE-461 and LE-462 silicone emulsion(s) and the remaining water. Add foaming agents, perfumes, etc. as desired.

Spray-and-wipe formulations should be packed in extruded, enameled, tin plate cans. Use a "Precision" valve with 0.080 in. nylon body, 0.030-in. nylon stem, buna-N gasket, and a "Precision" reverse-taper mechanical breakup actuator (0.016-in.). Aerosol with 5% by weight isobutane propellant or 7 to 10% by weight of a 60/40 mix of "Ucon" 114/12 Propellant.

For use, spray light film on surface or on dust cloth, and wipe with circular motion until dry. Avoid over-application. (Polish film is slippery; therefore, use caution to prevent over-spray from falling on floor or stairs.)

### Liquid Solvent Polishes

| | No. 1 | No. 2 | No. 3 | No. 4 | No. 5 | No. 6 | No. 7 |
|---|---|---|---|---|---|---|---|
| "Duroxon" E-321 | — | — | 5.0 | — | — | — | — |
| "Duroxon" E-421R | — | — | 5.0 | — | — | — | — |
| "Duroxon" E-421R | — | — | — | 8.0 | — | — | 8.0 |
| FT-150 | — | — | — | — | — | 4.0 | — |
| FT-300 | — | 1.6 | — | — | 1.5 | — | — |
| "Duroxon" J-324 | 4.0 | — | — | 2.0 | — | — | — |
| "Duroxon" R-21 | — | 1.6 | — | — | 1.5 | 4.0 | 4.5 |
| GS Wax Vi-291 | — | — | — | — | — | 8.5 | — |

| | | | | | | | |
|---|---|---|---|---|---|---|---|
| "Duron" ozokerite 71/74 | 4.0 | — | 3.0 | — | — | — | — |
| "Duron" 180 natural | — | — | — | — | — | — | 4.0 |
| "AC"-6 polyethylene | — | — | 1.0 | — | — | — | — |
| "Aristowax" 165 | — | 8.8 | — | — | 7.5 | — | — |
| Fully refined paraffin 143/150 AMP | — | — | — | — | — | 4.0 | — |
| Fully refined paraffin 130/132 AMP | 4.0 | — | — | — | — | — | 8.5 |
| Fully refined paraffin 125/127 AMP | — | — | — | 5.0 | — | 4.5 | — |
| "Hydrofol" glycerides T-57-N | — | — | — | — | 1.5 | — | — |
| "Cab-O-Sil" M-7 | — | — | 1.5 | — | — | — | — |
| "Amsco" 46 mineral spirit | 88.0 | 88.0 | 89.5 | 85.0 | 88.0 | 75.0 | 75.0 |
| Solids content (%) | 12.0 | 12.0 | 10.5 | 15.0 | 12.0 | 25.0 | 25.0 |

Description of Products

No. 1: Low-solids lquid paste wax.

No. 2: Snow-white, opaque, low cost, readily buffable product.

No. 3: Antiskid colloidal. Silica content polish for wood and terrazzo floors.

No. 4: General-purpose liquid solvent wax, 15% solids.

No. 5: Liquid solvent wax to meet U.S. Government requirements, Specification PW-158.

No. 6: Terrazzo seal, high-gloss concentrate.

No. 7: Terrazzo and wood seal and polish concentrate.

---

**Solvent Wax Polishes for U.S. Government**

Specification PW-158D

Wax Polish, General-Purpose, Solvent Type, Liquid

| | |
|---|---|
| "Duroxon" R-21 | 1.5 |
| "FT"-WAX 300 | 1.5 |
| "Aristowax" 165 | 7.5 |
| "Hydrofol" glycerides T-57-N | 1.5 |
| "Amsco" 46 mineral spirits or equivalent | 97.0 |

Wax Polish, General Purpose, Solvent Type, Liquid with Silica

| | |
|---|---|
| "Duroxon" E-321 | 5.0 |

| | |
|---|---|
| "Duron" Ozokerite 71/74 | 3.0 |
| "AC"-6 polyethylene | 1.0 |
| "Cab-O-Sil" M-7 | 1.5 |
| "Amsco" 46 mineral spirits or equivalent | 96.5 |

**Liquid Solvent Polish**

(May also be aerosoled)

| | |
|---|---|
| "Dow Corning" 531 fluid (as supplied) | 4.50 |
| "Dow Corning" 530 fluid | 0.75 |
| Wax | 2.00 |
| Stoddard solvent | 10.00 |
| VM & P naphtha | 82.75 |

Mix the "Dow Corning" 530 fluid, "Dow Corning" 531 fluid, wax and Stoddard solvent. Heat until the wax melts (caution inflammable). Add naphtha with stirring. (If desirable, the naphtha may be warmed to reduce wax precipitation.) Cool with stirring. Package.

Isopropyl alcohol (absolute) may be added in a small amount to improve solubility of the silicone in the solvent. Kerosene may also be added to slow evaporation rate. Abrasives may be added to the system as needed to give better cleaning.

**Aerosol Glaze**

| | |
|---|---|
| Liquid solvent polish (above) | 50 |
| Propellent 11 | 15 |
| Propellent 12 | 35 |

The selection and dispersion of the wax should be adequate so there is no valve clogging. Free halide in chlornated or fluorinated solvents may cause slight haziness upon reacting with the amine. This slight haziness, if observed, does not reduce the performance of the polish.

**Aerosol Polish**

For Leather, Metal, Wood and Plastics

| | |
|---|---|
| "Duroxon" J-324 | 1.5 |
| "Duroxon" B-120 | 0.5 |
| Silicone oil, 1000 cstks. | 1.0 |
| Silicone oil, 30,000 cstks. | 0.5 |
| "Amsco" Mintrol spirits or equivalent | 10.0 |
| VM & P Naptha | 86.5 |

Aerosol Fill

50% above concentrate
15% Propellant 11
35% Propellant 12

## Liquid Silver Polish

No. 1

| | | |
|---|---|---|
| A. | "Veegum" | 2.00 |
| | Sodium carboxymethylcellulose (7LP) | 0.10 |
| | Water | 76.90 |
| B. | Hydrated amorphous silica | 15.00 |
| C. | Buffer solution* | q.s. |
| D. | "Triton" X-102 | 5.00 |
| E. | Octadecyl thioglycolate | 1.00 |
| | Color | q.s. |
| | Perfume | q.s. |
| | Preservative | q.s. |

*Buffer Solution: 60 parts 1M citric acid, saturated sodium
40 parts citrate sol. (87 g/100 cc) filtered.

1. Dry-blend the "Veegum" and the CMC. Add to the water slowly, continually agitating until smooth.
2. Add B to 1. Buffer to pH 5.0 and add D.
3. Add E to 2 and mix until well dispersed.

For use, pour small amount of cleaner onto a damp cloth and clean article with moderate rubbing. Rinse article with water, dry and polish with a clean dry cloth. This formula is illustrative of the type of silver polish that prevents tarnish for several months. It has good resistance to hydrogen sulfide and ultraviolet light.

No. 2

| | |
|---|---|
| Infusorial earth or precipitated chalk | 48 |
| Diglycol stearate | 7 |
| Soda ash | 1 |
| Trisodium phosphate | 1 |
| Water | 70 |

The water and stearate are heated to 150°F and stirred until homogeneous. Then, the other ingredients are added and mixed to a smooth paste. Perfume is added as required.

The pink pastes usually have added jewelers' red rouge powder.

## Shoe Polish (Black)

| | | |
|---|---|---|
| A. | "Castorwax" | 4.2 |
| | Carnauba wax | 1.4 |
| | Crude montan wax | 5.0 |
| B. | Stearic acid | 2.0 |
| | Nigrosine base | 1.0 |
| C. | Ceresin | 13.0 |
| | "Paricin" 9 | 2.5 |
| D. | Turpentine | 90.0 |

Melt A together in a water-heated kettle. Melt and dissolve B separately. Combine A and B. Stir in the ceresin and "Paricin" 9 (C). When uniformly melted and mixed, eliminate all flames and add the turpentine

(D) with stirring. Cool mix to 105°F and pour into suitable containers. Let stand undisturbed overnight.

### Aerosol Auto Polish (Foam Type)

Concentrate

| | |
|---|---|
| Part I | |
| Silicone oil 1,000 cstks. | 5.0 |
| Silicone oil 30,000 cstks. | 1.0 |
| "Duroxon" J-324 | 2.5 |
| "Durmont" E or ES | 2.0 |
| "Amsco" Mintrol spirits | 12.0 |
| Kerosene | 8.0 |
| Part II | |
| Triethanolamine | 2.5 |
| Diethyl amino ethanol | 4.0 |
| Water | 16.0 |
| "Snowfloss" | 14.0 |
| "Methocel" 4,000 cps, 1% aqueous sol. | 37.0 |

Aerosol Fill

| | |
|---|---|
| Concentrate 35B-33B or 33F | 90.0 g |
| Propellant 12 | 10.0 g |

### High-Gloss Auto Polish Cream

The formula described in the following yields a liquid, creamy auto polish for packaging in plastic squeeze bottles or glass bottles. The product is easy to apply and buffs to a high gloss with very little effort.

| | | |
|---|---|---|
| 1. | "Duroxon" B-120 | 2.00 |
| | "Efton" UC regular | 3.00 |
| | "Arlacel" C | 1.50 |
| | Silicone fluid SF-96/1000 ctks. | 2.50 |
| | "Amsco" mineral spirits | 16.00 |
| | "Amsco" odorless 450 solvent | 4.00 |
| 2. | Water at 190°F | 69.75 |
| | "Carbopol" 934 | 0.25 |
| 3. | Sodium hydroxide, 10% aqueous solution | 1.00 |

Dissolve the "Duroxon", "Efton", "Arlacel" C and silicone fluid in the solvents at 190 to 200°F. Disperse the "Carbopol" in the water in a separate container at 190 to 200°F with high-shear agitation. Pour

1 into 2 with agitation. Add 3 and continue high-speed agitation to max. 100°F or room temperature.

---

### Automobile Paste Polish

This is completely free of any abrasives which might damage the modern types of acrylic car finishes. It is intended primarily for the new-car dealer and similar institutional uses, where the painted surface is treated with a cleaner first, to remove stains and road film. The polish gives excellent, smear-free jetness on dark colors, and is highly water-resistant. Modifications are easily accomplished by incorporating mild abrasives, cleaners or higher-boiling solvents (the latter, when the product is intended for use in hot climates or under direct sun).

| | |
|---|---|
| "Duroxon" J-324 or J-324 V | 26.5 |
| "Mycrox" 114 | 6.7 |
| Silicone fluid 350 cstks. | 6.8 |
| "Amsco" 46 mineral spirit | 60.0 |

Melt the "Duroxon" and the "Mycrox". Preheat the "Amsco" slightly and add the silicone fluid. Slowly incorporate the silicone solution with the solvent into the wax melt. Cool with agitation to about 138°F, then pour the melt, all at once, into suitable tins or jars. Allow to cool undisturbed. Cover the containers when cool.

---

### Paste Automobile Cleaner-Polish

This formula yields a medium hard consistency automotive paste wax for the comsumer trade with exceptionally good application, spreading, and rub-out properties. The gloss and the depth of the gloss are rated as excellent, and so is the durability and water-resistance of the film.

| | |
|---|---|
| "Durmont" E | 11.0 |
| "Mycrox" 114 | 7.0 |
| "Duron" 195-D | 11.0 |
| Silicone oil (1,000 cstks) | 7.5 |
| Silicone oil (30,000 cstks) | 16.5 |
| Kerosene | 62.5 |
| "Amsco" Mintrol spirits | 46.0 |
| 1% solution of BMA-6202 Dye in "Amsco" Mintrol spirits | 2.5 |
| "Snowfloss" | 12.0 |
| "Kaopolite" SFO | 12.0 |

| | |
|---|---|
| "Satintone" Clay | 12.0 |

Temperature: Approximately 160°F.

**Auto Wax-Wash**

| | |
|---|---|
| Silicone fluids | 20.0 |
| "Surfactol" 318 | 9.8 |
| Paraffin oil | 7.5 |
| Triethanolamine | 1.2 |
| Ammonia, 26° Be | 0.7 |
| Water | 60.8 |

**Vinyl Auto Top Polish**

| | No. 1 | No. 2 |
|---|---|---|
| "Efton" D Super | 50.0 | 50.0 |
| "AC"-540 | 50.0 | 50.0 |
| Oleic acid | 3.0 | 3.0 |
| KOH at 45% | 5.0 | 5.0 |
| Diethyl amino ethanol | 14.0 | 14.0 |
| Methyl "Carbitol" | — | 6.0 |
| Water | 299.0 | 293.0 |

**Soft Polishing Wax Paste**

This formula results in a smooth white creampaste with excellent application properties and gloss on wood and "Formica". It can be used for shoe polishes (clear or dyed), auto polishes furniture creams paste floor polishes kitchen-cabinet waxes, and plastic-mold release agents. The distinguishing feature of this system is the incorporation of a high percentage of water in a solvent polish without the presence of any emulsifier.

| | |
|---|---|
| "Duroxon" H-111 | 8 |
| "Durmont" 500 | 3 |
| "Duroxon" E-321 | 2 |
| Fully refined paraffin wax (141°F) | 10 |
| "Amsco" mineral spirits | 25 |
| Water | 52 |

Dissolve the waxes in the mineral spirits by heating to 200°F. Reduce the temperature of the solution to 180°F while agitating. Add the water at 180°F with stirring. Continue stirring while reducing the temperature of the mixture to 120°F. At 120°F pour the mixture into suitable cans and allow to stand undisturbed for good paste formation.

To make black shoe polish, add to the above 2 parts "Excelsior" black and 0.5 calco nigrosine base Z-1630, which should first be dissolved in an equal quantity of stearic acid (double-pressed).

---

**Chrome Protective Wax**

(Aerosol)

| | |
|---|---|
| "Ruhrwax" K-20 | 15 |
| Mineral spirits | 30 |
| Fast-evaporating mineral spirit | 55 |

Filling

150 ml chrome protective wax solution, 50 ml propellant 1P/114 (1:1).

Chapter XVI

# PYROTECHNICS

### Safety Match

No. 1

| | |
|---|---|
| Animal glue | 10 |
| Starch | 2-3 |
| Sulfur | 3-5 |
| Potassium chlorate | 45-55 |
| ZnO, $CaCO_3$ | 3 |
| Diatomaceous earth | 5-6 |
| Powdered glass or "fine" silica | 15-32 |
| $K_2Cr_2O_7$ or $PbS_2O_3$ | to suit |
| Water-soluble dye | to suit |

The best striking qualities are obtained by use of very high grade glue (such as Peter Cooper "Grade IIa Extra" with foaming properties), leading to a match paste of d=1.20 to 1.35 at a water content of 30 to 32%.

No. 2

| | |
|---|---|
| Animal glue | 11 |
| Sulfur | 5 |
| Potassium chlorate | 51 |
| Zinc oxide | 7 |
| Black iron oxide | 6 |
| Manganese dioxide | 4 |
| Powdered glass | 15 |
| Potassium dichromate | 1 |

*These formulas are self-igniting and explosive. If you have had no experience in this field start with very small amounts and use caution.

### Sparklers

| | No. 1 | No. 2 |
|---|---|---|
| Potassium perchlorate | 60.0 | — |
| Barium nitrate | — | 50.0 |
| Potassium chlorate | — | — |
| Aluminum, dark pyro | 30.0 | 8.0 |
| Dextrin | 10.0 | 10.0 |
| Steel filings | — | 30.0 |
| Charcoal fine | — | 0.5 |
| Neutralizer | — | 1.5 |

The outside of the composition is coated with magnesium-aluminum alloy grit.

---

**Rain Cloud-Seeding Crystals**

No. 1

| | |
|---|---|
| Silver iodide or lead iodide | 40-60 |
| Ammonium perchlorate | 24-45 |
| Synthetic resin Binder | 10-25 |
| Graphite or oil | 1.5-2 |

| | No. 2 | No. 3 |
|---|---|---|
| Silver iodate | 75.0 | — |
| Lead iodate | — | 75.0 |
| Magnesium 25$\mu$ | 15.0 | 10.0 |
| "Laminac" | 10.0 | 15.0 |

Prior to addition of the resin, the components are moistened with acetone and the mixture pressed into grains several feet long and four inches in diameter.

---

**Strike-Anywhere Match Tip and Base Compositions**

| | Tip | Striking Base |
|---|---|---|
| Animal glue | 11.0 | 12.0 |
| Starch | 4.0 | 5.0 |
| Paraffin wax | — | 2.0 |
| Potassium chlorate | 32.0 | 37.0 |
| Phosphorus sesquisulfide | 10.0 | 3.0 |
| Sulfur | — | 6.0 |
| Rosin | 4.0 | 6.0 |
| Dammar gum | — | 3.0 |
| Infusorial earth | — | 3.0 |
| Powdered glass and other filler | 33.0 | 21.5 |
| Potassium dichromate | — | 0.5 |
| Zinc oxide | 6.0 | 1.0 |

---

**Underwater Flare**

| | |
|---|---|
| Linseed oil | 8 |
| Manganese dioxide | 1 |
| Magnesium, powdered | 16 |
| Aluminum, powdered | 12 |
| Barium sulfate | 40 |
| Barium nitrate | 32 |

**Red Signal Flare**

| | No. 1 | No. 2 | No. 3 |
|---|---|---|---|
| Magnesium powder | 21.0 | 17.5 | 30.0 |
| Strontium nitrate | 45.0 | 45.0 | 42.0 |
| Potassium perchlorate | 15.0 | 25.0 | 9.0 |
| Hexachlorobenzene | 12.0 | — | — |
| Polyvinylchloride | — | 5.0 | 12.0 |
| Gilsonite | 7.0 | 7.5 | — |
| "Laminac" | — | — | 7.0 |

**Black Gunpowder**

| | No. 1 | No. 2 | No. 3 |
|---|---|---|---|
| Charcoal JAN-C-178A | 15.6 | — | 16.0 |
| Semibituminous coal | — | 14.0 | — |
| Sulfur JAN-S-487 | 10.4 | 16.0 | 12.0 |
| Potassium nitrate MIL-P-156B | 74.0 | 70.0 | — |
| Sodium nitrate MIL-S-322B | — | — | 72.0 |

**Railroad Torpedo**

| | |
|---|---|
| Potassium chlorate | 40 |
| Sulfur | 16 |
| Sand (60-mesh) | 37 |
| Binder | 5 |
| Calcium carbonate | 2 |

**Oxygen Candle**

No. 1

| | |
|---|---|
| Sodium chlorate | 92 |
| Barium peroxide | 4 |
| Steel wool, grade 2 or 00 | 4 |

No. 2

| | |
|---|---|
| Lithium perchlorate | 84.82 |
| Lithium peroxide | 4.24 |
| Manganese metal powder | 10.94 |

**Black Powder, Military**

| | No. 1 | No. 2 |
|---|---|---|
| Pot. Nitrate | 74.0 | — |
| Sod. nitrate | — | 71.0 |
| Charcoal, powdered | 15.6 | 16.5 |
| Sulfur | 10.4 | 12.5 |

The nitrate and other materials are ground and sieved separately. They are then dampened and mixed carefully, to evaporate moisture. Great caution must be used to prevent explosion.

Chapter XVII

# RUBBER, PLASTICS, WAXES

## Vinyl Floor Tile

| | No. 1 | No. 2 | No. 3 |
|---|---|---|---|
| PVC polymer | 100 | 100 | 100 |
| DOP | 35 | — | — |
| "Cereclor" S.52 | — | 44 | 51 |
| Ground limestone | 350 | 350 | 362 |
| Asbestos | 100 | 100 | 108 |
| Barium zinc stabilizer | 3 | 3 | 3 |
| Epoxy stabilizer | 3 | 3 | 3 |
| Titanium dioxide | 5 | 5 | 5 |
| Colorant | 2 | 2 | 2 |

No. 4

| | |
|---|---|
| Polyvinyl chloride resin | 100.00 |
| Calcium carbonate fitter | 150.00 |
| Clay filler | 75.00 |
| Titanium dioxide | 8.75 |
| Colored pigments | 1-2.00 |
| Ba-Cd-Zn stabilizer | 2.50 |
| Epoxidized soybean oil | 5.00 |
| Plasticizer | 34.00 |

| | No. 5 | No. 6 | No. 7 | No. 8 |
|---|---|---|---|---|
| PVC low fusion | 100.00 | 100.00 | 100.00 | 100.00 |
| DIOP | 40.00 | — | — | — |
| Butyl benzyl phthalate | — | 40.00 | — | — |
| Tricresyl phosphate | — | — | 40.00 | — |
| "Cereclor" S.52 | — | — | — | 50.00 |

| | | | | |
|---|---|---|---|---|
| Ground limestone | 200.00 | 200.00 | 200.00 | 200.00 |
| Ba/Cd stabilizer | 2.00 | 2.00 | 2.00 | 2.00 |
| Epoxy stabilizer | 3.00 | 3.00 | 3.00 | 3.00 |
| Calcium stearate | 1.00 | 1.00 | 1.00 | 1.00 |
| Titanium dioxide | 6.00 | 6.00 | 6.00 | 6.00 |
| Carbon black | 0.03 | 0.03 | 0.03 | 0.03 |

**Vinyl Asbestos Tiles**

| | No. 1 | No. 2 | No. 3 | No. 4 |
|---|---|---|---|---|
| VC/CA copolymer (16% acetate) | 100 | 100 | 100 | 100 |
| DIOP | 40 | — | — | — |
| Butyl benzyl phthalate | — | 40 | — | — |
| Tricresyl phosphate | — | — | 40 | — |
| "Cereclor" S.52 | — | — | — | 50 |
| Ground limestone | 250 | 250 | 250 | 250 |
| Asbestos | 250 | 250 | 250 | 250 |
| Ba/Zn stabilizer | 3 | 3 | 3 | 3 |
| Calcium stearate | 1 | 1 | 1 | 1 |
| Titanium dioxide | 20 | 20 | 20 | 20 |

**Plastisol Sheeting**

| | |
|---|---|
| "Geon" 121 restin | 55 |
| "Geon" 202 resin | 45 |
| Ba-Cd-Zn stabilizer | 3 |
| Epoxidized soybean oil | 3 |
| Calcium carbonate | 10 |
| Plasticizer | * |

*As required to produce Brookfield viscosity of 4000 cp at 23°C (No. 4 spindle, 6 rpm).

**PVC Extrusion Upholstery Welting**

| | |
|---|---|
| "Geon" 101 | 100.0 |
| "Atomite" | 10.0 |
| Ba-Cd-Zn stabilizer | 3.0 |
| Expoxidized soybean oil | 3.0 |
| "Epolene" E-11 resin | 0.5 |
| Plasticizer 466 | 42.0 |

**Antifogging PVC Film**

| | |
|---|---|
| PVC (Dow resin 144) | 100 |
| Dioctyl phthalate | 20 |
| Epoxidized soybean oil | 15 |
| Zinc stearate | 5 |
| "Span" 60 | 1 |
| "Tween" 60 | 1 |

**Fluorescent Cast Vinyl Film**

| | |
|---|---|
| "Flexol" R2H plasticizer | 24.1 |
| "QYNV" resin | 34.6 |
| "Mark" LL | 1.0 |
| G-62 epoxy plasticizer | 1.8 |
| Pigment P-1600 | 24.5 |
| "Socal" 3 solvent | |

| | |
|---|---|
| (95% aromatic) | 14.0 |

Disperse all "QYNV" in R2H until a uniform paste is formed. Then add the pigment gradually until all is dispersed well. Add G-62 and Mark LL and finally the solvent. Do not overheat during dispersion.

Note that the paint thickens up in about 1 or 2 weeks on standing. Therefore more solvent should be added to bring the paint to the cating viscosity if it is to be used at a later date.

---

**PVC-ABS Blend**

| | |
|---|---|
| ABS resin | 0-25% |
| PVC resin | 100-75% |
| Liquid "Organo" tin mercaptide | 2 parts |
| Magnesium stearate | 1 part |

---

**Polyester Emulsion Casting Compound**

No. 1

| | |
|---|---|
| "Aropol" WEP26 | 400.0 |
| Cobalt octoate (12%) | 5.0 |
| Dimethyltoluidine | 2.0 |
| Water | 600.0 |
| "Lupersol" DSW | 2.0 |

No. 2

| | |
|---|---|
| "Aropol" WEP26 | 400.0 |
| "Lucidol" 70 (wetted BPO) | 5.6 |
| Water | 600.0 |
| Dimethyltoluidine | 1.2 |

---

**Polyester Premix Molding**

| | |
|---|---|
| "Paraplex" P19A (60 parts P340/40 parts P543) | 35 |
| "ASP"-400 (clay) | 35 |
| Asbestos 7T-15 | 5 |
| -in H.S.I. glass fibers | 25 |
| tert-Butyl peroctoate | 1.0% based on P19A |
| "Zelec" UN | 0.5% based on P19A |

Alternate catalysts: tert-Butyl perbenzoate, benzoyl peroxide. Mixing times: 2 min for resin and fillers, 3 min after addition of glass.

"Zelec" UN is the recommended release agent. Apart from these components, the formulation may be modified as desired to produce optimum-quality parts. After the formulation has been mixed, parts should be molded promptly. If there is to be a delay before molding, the material should be wrapped in cellophane, polyethylene film, or wax paper.

**Reverse Roll, Knife and Gravure Fluorescent Coating**

Solvent System

| Disperse: | |
|---|---|
| "Pliolite" S-5D, butadiene-styrene 40% in xyol diluted with mineral spirits 50% | 15.0 |
| "Pliolite" S-7, butadiene-styrene 30% in toluene | 29.0 |
| Resin 276-V2 | 2.0 |
| "V.G." pigment R-103-G or R-203-G | 35.0 |
| Thin with mineral spirit (low-boiling) | 19.0 |

Water Emulsion System

| Disperse: | |
|---|---|
| $H_2O$ | 20.6 |
| "Tamol" 731 dispersant (25%) | 0.8 |
| "Cascoloid" ST-40 wetting agent (50%) | 1.4 |
| Antifoamer | 0.2 |
| "Velva-Glo" pigment R-103-G or R-203-G | 34.0 |
| Mix with AC-33, acrylic emulsion | 43.0 |

Fluorescent Vinyl Plastsol and Organosol for Coating

| | |
|---|---|
| Vinyl resin | 100 |
| Plasticizer | 50-80 |
| "V.G." pigment | up to 80 |
| Epoxy stabilizer | 5-10 |
| Barium-cadmium stabilizer | 2-3 |

Plastisols or organosols are prepared by mixing the ingredients in a heavy-duty mixer having higher shear rather than high speed. The vinyl resin is added in portions to a mixture of plasticizers and stabilizers, with mixing continued until a uniform paste is obtained. Thereafter the pigment is added in portions and folded quickly in the plastisol with slow stirring until a smooth paste, free of agglomerates, is obtained. Deaeration, if necessary, should be carried out at this point.

With high pigment loads and/or nonmigrating plasticizers addition of solvent may be desirable to obtain the proper viscosity for coating (organosol). Mineral spirits, aromatics, DIBK or mixtures thereof are useful. A minimum of solvent should be used; solvent addition represents the last step in the organosol preparation.

The following formulation has been successful for the preparation of a 3 to 4-mil cast vinyl film of exceptional fluorescent brightness.

| | |
|---|---|
| Vinyl resin QYNV | 39.7 |

| | |
|---|---|
| Dioctylphthalate | 11.9 |
| Dioctyladipate | 11.9 |
| "Paraplex" G-62 | 2.0 |
| "Mark LL" (Ba-Cd stabilizer) | 1.2 |
| "V.G." pigment, R-103-G or R-203-G | 25.3 |
| "Socal" 2 | 4.0 |
| "Socal" 3 | 4.0 |

---

**Extrusion Compounds**

| | Opaque | Clear | Cable insulation | Cable sheathing and general purpose |
|---|---|---|---|---|
| Vinyl polymer | 100 | 100 | 100 | 100 |
| DIOP or DOP | 50 | 37 | 34 | 42 |
| "Cereclor" 42 | 25 | 20 | 17 | 21 |
| Filler | 30 | — | — | — |
| White lead | 5 | — | — | 7 |
| Tribasic lead sulfate | — | — | 5 | — |
| Ba/Cd/Zn system+ chelating agent | — | 2 | — | — |
| Lubricant | 1 | 0.5 | 1 | 1 |
| Electrical clay | — | — | 10 | — |
| Pigments | as required | | | |

---

**Plastisols for Spread Coating**

| | High-Viscosity | Low-Viscosity |
|---|---|---|
| Plastisol-making polymer | 100 | — |
| Plastisol-making polymer | — | 100 |
| DIOP | 66 | 34 |
| TCP | — | 34 |
| "Cereclor" 42 | 33 | 34 |
| Inorganic filler | 40 | — |
| Ba/Cd stabilizer | — | 2 |
| Epoxidized oil | — | 3 |
| White lead | 4 | — |
| Pigment | as required | |

---

**9—Medium Viscosity Plastisols for Casting**

| | |
|---|---|
| Plastisol-making polymer | 100 |
| DIOP or DOP | 59 |

| | |
|---|---|
| "Cereclor" 42 | 31 |
| Ba/Cd stabilizer | 2 |
| Epoxidized oil | 3 |
| Pigment | as required |

This formulation is suitable for toys. Calcium stearate (up to 2 parts) can be used alone for slush molding applications but should be reinforced with an epoxy compound for more rigorous operations.

---

**Brattice Cloth Plastic Coating**

| | |
|---|---|
| Plastisol-making polymer | 50 |
| Granular polymer | 50 |
| TCP | 70 |
| "Cereclor" 42 | 50 |
| Whiting | 120 |
| Antimony oxide | 10 |
| White lead | 3 |

The high filler content of this formulation permits the use of a higher "Cereclor" 42/plasticizer ratio than that recommended for general use.

---

**Non-Inflammable Conveyor Belting**

| | Carcass | Cover |
|---|---|---|
| Plastisol-making polymer | 40 | 40 |
| Plastisol-making polymer | 60 | 60 |
| TCP | 44 | 46 |
| DOP | 20 | 10 |
| "Cereclor" 42 | 32 | 28 |
| Calcium stearate | 2 | 2 |
| Lead silicate | 2 | 2 |

---

**"Geon" Organosol**

| | |
|---|---|
| "Geon" 121 resin | 100 |
| Ba-Cd-Zn stabilizer | 3 |
| DOP | 25 |
| "Kodaflex" TXIB | 25 |

Aliphatic hydrocarbon solvent (B.P. 156-198°C., aniline pt. 62°C.) to suit.

---

**Polyurethane Foams**

As an emulsifier in flexible urethane foams, "Surfactol" 365 (0.1%-5.0% on the weight of the polyol used) serves to emulsify the catalyst and the water portion of the formulation, prior to mixing with ths polyol and diisocyanate in the foaming machine. The use of "Surfactol" 365 prevents the formation of "hot spots" during foaming due to incomplete emulsification of the water. Being an hydroxyl compound, "Surfactol" 365 also enters into the urethane reaction by chemical cross-linking.

### Polyurethane Potting Compound

| | | |
|---|---|---|
| Part A | "Polycin" U-63 | 60.00 |
| Part B | "Polycin" 12 | 40.00 |
| | Triethylamine Diamine | 0.03 |

### Epoxy Potting Compound Topping

| | | |
|---|---|---|
| Base resin | "ERL" 2795 | 100.00 |
| | "Epotul" 6130 | |
| Liquid polyamide | "Versamid" 125 | 7.00 |
| Class 5 harderner | "ERL"—2807 | 15.00 |
| | "Epi-Cure" 87 | |
| Pigment | Red paste Px-2523 | 0.25 |
| | Piper red 37-T-58 | |
| Antifoaming agent | Antifoam "A" | 0.01 |

### Epoxy Casting Resins

No. 1

| | |
|---|---|
| "Lekutherm" X 100 | = 100 |
| "Lekutherm" hardener H | = 100 |
| "Desmorapid" DB | = 2 |

Cure: 4 hr at 80°C+16 hr at 120°C.

No. 2

| | |
|---|---|
| "Lekutherm" X 100 | = 100 |
| "Lekutherm" hardener H | = 100 |
| "Desmorapid" DB | = 2 |
| Quartz flour | = 400 |

Cure: 4 hr at 80°C+16 hr at 120°C.

### Flame Retardent Polypropylene

| | |
|---|---|
| Polypropylene | 73.8 |
| "Chlorowax" 70 | 12.5 |
| Antimony trioxide | 12.5 |
| "Tribase" E | 1.0 |
| "Akroflex" C | 0.2 |

### Plastics Processing Aids

Phenolic Resins

"Castorwax" added at a level of from 0.5 to 1.0% to phenolic resins, after manufacture but before grinding, acts initially as a grinding aid and subsequently as an antiblocking agent, internal lubricant and mold release during molding operations.

Polyethylene

"Castorwax" and "Castorwax" F-1 added at a level of from 2 to 3% to low-molecular-weight polyethylene improves the processing and subsequent grease-resistance of the sheeted polymer. "Castorwax" used at a level of 0.5% improves the molding properties of reworked and pelletized polyethylene and acts as a mutual solvent for rendering ethylcellulose compatible with polyethylene.

Polypropylene

"Paricin" 285 at a level of 0.1 parts per hundred of polypropylene reduces the friction and static build-up of this resin to permit easier processing and production of sheeted goods.

Polystyrene

"Paricin" 285 used at a level of ·/8 of 1% improves the processing characteristics of high-impact styrene. It also improves the overall appearance and color development of extruded and molded parts.

Rigid Polyvinylchloride

"Paricin" 220 and "Paricin" 285 used up to a level of 2% act as internal lubricants for rigid vinyls. "Paricin" 285 is recommended as an extrusion lubricant while "Paricin" 220 is recommended for calendering.

---

**High-Temperature-Resistant Resin**

| | |
|---|---|
| "Dexsil" 201 polymer (15 g) | 100.0 |
| "Mapico" red No. 297 (1.5 g) | 10.0 |
| "Min-U-Sil" (5 micron) (12.0 g) | 80.0 |
| "Luperco" CST (0.23 g) | 1.5 |

Compounding

1. Starting with the mill at room temperature, pass the "Dexsil" 201 crumb through the mill until banded.
2. Work the stock until masticated and a smooth sheet is obtained.
3. Add the iron oxide (Mapico Red or RO 3097) antioxidant across the crepe.
4. Add any antioxidant that has fallen through the mill.
5. Mill the stock until the iron oxide is evenly dispersed.
6. Add the Min-U-Sil filler slowly and evenly across the crepe.
7. Add any filler that has fallen through the mill.
8. By this time the stock may be adhering to the rear roll and if all the Min-U-Sil has not been added the remainder may be added at this time.

9. Work the stock until the filler is well dispersed.
10. Add the Luperco CST curing agent slowly and evenly across the bank.
11. Work the stock sufficiently to disperse the curing agent.

**Note:** If the compound is worked too long after the addition of curing agent, there is a tendency for the material to stick to the mill. This should be avoided. If sticking occurs, opening the bite of the mill may alleviate this problem.

12. Opne the rolls and sheet out the compounded stock.

Resting

13. Hold the compounded stock at room temperature for 72 hours.

Molding

14. Prior to molding freshen the stock on the mill and sheet out smoothly to 1/10″ thickness.
15. Cut a preform from the sheet slightly smaller than the 2″×5″ mold.
16. Treat the mold with "Mc-Lube" No. 1700 and heat to 230°F in a press.
17. Place the preform in the mold and slowly close the press.
18. Apply a pressure of 250-300 psi for 15 sec. Reduce to 50 psi for balance of cure.
19. Cure for 5 min at 230°F.
20. Release pressure and remove specimen from mold.

Postcure

21. Post cure the sample in a circulating air oven using the sequential schedule:
    16 hrs. at 212°F.
    8 hrs. at 300°F.
    24 hrs. at 500°F.

Physical Properties

A specimen prepared following the above procedure will have these typical physical properties:

Tensile strength 500-600 psi
Elongation 200-250%
Hardness (Shore A) 55-60

Service Temperature

600°F long exposure
900°F brief exposure

**Nylon Solvent**

| | No. 1 | No. 2 |
|---|---|---|
| Phenol | 85 | 50 |
| Methanol | 15 | — |
| Formic acid (95%) | — | 50 |

### Epoxy Ester

| | |
|---|---|
| "Epon" 1004 | 60 |
| Fatty acid ("Baker 9-11") | 40 |
| Xylene—to 45% N.V. after correction for loss due to eater formation | |

Charge the epoxy resin and fatty acid to a three-neck flask fitted with an agitator, thermometer, water trap, and reflux condenser; blanket with $CO_2$. Heat to 500°F and esterify using a xylene azeotrope until an acid number of 2 is obtained. Cool and reduce to 45% N.V. with xylene. After cooling to room temperature add 0.005% cobalt drier as metal (for baking) or 0.04% cobalt drier as metal (for air dry), based on total solids of the epoxy ester.

---

### Acrylic Polymer

| | |
|---|---|
| Water | 280 |
| "Abex" 185 surfactant | 36 |
| Ethyl acrylate | 175 |
| Methyl methacrylate | 50 |
| "Sipomer" HEM | 25 |
| Methacrylic acid | 2 |
| Potassium persulfate | 1 |
| Sodium bisulfite (2% solution) | 40 |

Charge water into reaction flask and purge with nitrogen for 15 min. Add "Abex" 18S and dissolve. Add 20% of premixed monomers—ethyl acrylate, methyl methacrylate, "Sipomer" HEM and methacrylic acid. Purge with nitrogen for another 5 min.

Heat contents of flask to 50°C and add potassium persulfate. To initiate the reaction add 10% of the sodium bisulfite solution. As soon as the reaction is started and after an exothermic temperature rise of 5°C is noticed, start adding the remainder of the monomers and bisulfite soltion, concurrently.

Maintain reaction temperature between 72 and 76°C throughout the addition period of the monomers. After addition is completed, heat the batch to 85°C. Maintain this temperature for a half hour. Then cool to room temperature and adjust pH to between 8 and 9.

---

### Vinyl Copolymer

| | |
|---|---|
| Water | 100.0 |
| "Abex" 18S | 14.0 |
| Vinyl acetate | 95.0 |
| "Sipomer" DMM | 5.0 |
| $K_2S_2O_8$ | 0.3 |
| $NaHSO_3$ (2% solution) | 6.0 |

It is suggested that the minimum batch size should be 500 g because of reaction rate considerations.

Charge water (1) into the reaction flask; purge with nitrogen for 15 min. Add "Abex" 18S (2), and allow to dissolve. Then add 20% of the premixed monomers, vinyl acetate (3) and "Sipomer" DMM (4), and emulsify. Purge the emulsion with nitrogen for five minutes. Heat the batch to 50°C before adding the $K_2S_2O_8$ (5); allow approximately one minute to dissolve. Then add one ml of the $NaHSO_3$ solution (6) to initiate the reactions.

After initiation, the heat of polymerization will raise the batch temperature. As the temperature reaches 60°C, start the flow of the remaining monomer mixture. During the addition time, allow the temperature to reach 70°C and maintain between 68 and 70°C throughout the reaction cycle. The total addition time for the monomer should require approximately 1 hr. During the monomer addition period, add incrementally one ml of bisulfite solution every 15 min. After addition of the monomers is completed, allow the temperature to rise to 80°C (external heating might be necessary). Maintain this temperature for one-half to one hour and continue the incremental addition of bisulfite solution. At the end of the heating cycle, the residual unreacted monomer will be less than 0.5%. The batch then can be cooled and removed from the reaction flask.

---

## Rosin Derivatives

### Limed Rosin

With heat and agitation, dissolve 100 parts rosin in 25 parts mineral spirits in a flask equipped with a stirrer, thermometer and condenser. Slowly add, at 130 to 140°C with stirring, a slurry containing 7 parts lime (6.5 parts lime for wood rosin), 14 parts mineral spirits and 0.25 parts glacial acetic acid; add in small increments to control foraming. Hold with agitation at 130 to 140°C for 30 to 60 min, then add sufficient mineral spirits for filtering (50% N.V. solution).

Filter and recover by steam distillation. An inert gas sparge can be used to help remove solvent until steam is introduced at 200°C. Continue up to 250°C and until distillate contains less than 10% "oils." Reintroduce $N_2$ sparge and cool below 220°C; hold to allow gas bubbles to dissipate and pour.

### Maleic Adduct

Charge 100 parts rosin and 5 parts maleic anhydride to a flask equipped with a stirrer, thermometer, condenser, and inert gas ($N_2$) sparge. Heat under nitrogen to 225°C with caution, as the reaction is exothermic. Check for free maleic by dimethylaniline spot test, and when none present, cool below 200°C and pour.

Glycerol Ester

To a standard resin kettle, equipped with stirrer, condenser with water trap, thermometer, and inert gas ($N_2$) sparge, charge 100 parts rosin and 9.6 parts glycerol. Heat under nitrogen to 200°C and hold for 15 min; gradually raise temperature to 275°C (over 1 hr), and hold for desired acid value. Remove last traces of water of esterification by vacuum or by immersing sparge tube after acid number is below 20.

Pentaerythritol Ester

Charge 100 parts of rosin to a resin kettle and heat to 200°C under nitrogen. Add 11.2 parts pentaerythritol (technical) and raise temperature to 295°C. Hold for desired acid value. Remove last traces of water of esterification by vacuum or by immersing sparge tube.

Maleic-Modified Pentaerythritol Ester

To a resin kettle, charge 100 parts of rosin and 11 parts maleic anhydride. Heat to 220°C over approximately 1 hr, and then add 0.25 parts calcium acetate and 16 parts pentaerythritol. Hold at 275°C for an acid value of 35 or less.

Phenolic-Modified Pentaerythritol Ester

First prepare a phenolic condensate by dissolving 5 parts sodium hydroxide in 50 parts water. Then add 73 parts Bisphenol-A and maintain at 55 to 60°C while adding 100 parts "Formalin". Hold at this temperature for 6 hr, then neutralize with 12.5 parts hydrochloric acid in 25 parts water. Wash to pH 5.5 and concentrate to desired solids content.

To complete the resin, charge 100 parts of rosin and 1 part maleic anhydride to a resin kettle. Heat to 190 to 200°C, and add 17.0 parts of phenolic condensate (100% solids basis). Hold at 200°C for 15 min, then add 0.25 parts calcium acetate, 7.0 parts glycerol and 7.0 parts pentaerythritol (technical). Raise temperature to 275°C and hold for an acid value of 12.20.

## SBR Extrusion Compound

| | |
|---|---|
| SBR 1821 | 217.5 |

| | |
|---|---|
| Zinc oxide | 5.0 |
| Clay (hard) + 2 parts water | 32.0 |
| PBNA | 2.0 |
| "Zenite" | 2.0 |
| "Thionex" | 0.6 |
| Sulfur | 2.5 |
| Paraffin wax | 2.0 |
| Stearic acid | 1.0 |
| "Desical" | 9.4 |
| "Circolite" process oil | 0.6 |

---

**Hard-Rubber Compounds**

| | High Grade Ebonite | Battery Case Stock | Grinding Wheel Stock |
|---|---|---|---|
| "Flosbrene" 25 HV | — | — | 100.0 |
| "Flosbrene" 25 VLV | 100.0 | 100.0 | — |
| Sulfur | 43.0 | 35.0 | 40.0 |
| Lime | — | 15.0 | 10.0 |
| Aldehyde-amine accelerator | 3.0 | — | 1.5 |
| Phenolic resin | — | — | 6.5 |
| "Armeen" HT | — | 1.5 | — |
| Petrolatum | 8.0 | — | — |
| Process oil | 10.0 | 25.0 | — |
| Hard rubber dust | 200.0 | — | — |
| Coal dust | — | 500.0 | — |
| "Alundum" (24F grit) | — | — | 1200.0 |
| Asbestos fiber | — | — | 5.0 |
| **Cure** | | | |
| Time | 20 min. | 6 min. | 60 min. |
| Temperature | 340°F | 365°F | 316°F |
| **Properties** | | | |
| Tensile, psi | 6000.0 | 2300.0 | — |
| Elongation, % | 3.5 | 2.0 | — |
| Shore D hardness | 80.0 | 80.0 | — |

---

**Synthetic Rubber Heel—Light Color**

| | |
|---|---|
| "Synpol" 8140 | 100.0 |

| | |
|---|---|
| Hard clay | 50.0 |
| "Silene" EF | 70.0 |
| "Solka Floc" | 15.0 |
| Process oil | 30.0 |
| "Cumar" MH 2½ | 10.0 |
| Zinc oxide | 5.0 |
| Stearic acid | 1.0 |
| Titanium dioxide | 10.0 |
| Color | (as necessary) |
| MBTS | 1.5 |
| DPG | 1.0 |
| Sulfur | 3.0 |

Cure 15 min at 315°F.

---

**Translucent Molded Soling**

| | |
|---|---|
| "Synpol" 8107 | 80.0 |
| "Synpol" 1551 | 20.0 |
| "Zalba" special | 1.0 |
| Zinc oxide | 1.0 |
| "Hi Sil" 233 | 60.0 |
| Light process oil | 35.0 |
| "Actisil" | 3.0 |
| "Monex" | 1.0 |
| "Zenite" A | 1.4 |
| Sulfur | 2.0 |

Cure for 10 min at 315°F.

---

**Sponge Arch Cushion**

| | |
|---|---|
| "Synpol" 8107 or 8140 | 100.00 |
| Whiting | 80.00 |
| Whiting | 80.00 |
| Zinc oxide | 3.00 |
| "Naugawhite" | 1.00 |
| Petrolatum | 7.00 |
| Paraffin wax | 1.50 |
| Paraffin oil | 40.00 |
| Bicarbonate | 10.00 |
| OXAF | 0.50 |
| "Delac"-S | 0.80 |
| Ethyl "Tuex" | 1.20 |
| Red iron oxide | 0.33 |
| Tan iron oxide | 1.00 |
| Sulfur | 3.50 |

Cure for 5 min at 320°F.

---

**Soft Cellular Soling**

| | |
|---|---|
| "Synpol" 8140 | 125.0 |
| "Cumar" MH 2½ | 10.0 |
| "Hi Sil" 233 | 30.0 |
| Hard clay | 75.0 |
| Zinc oxide | 5.0 |
| Stearic acid | 5.0 |
| Petrolatum | 13.0 |
| Light process oil | 20.0 |
| Diethylene glycol | 1.0 |
| "Naugawhite" | 1.0 |
| "Celogen" | 7.0 |
| MBTS | 1.5 |
| "Tuex" | 1.5 |
| DPG | 0.6 |
| Sulfur | 3.5 |

Cure 7.5 min at 324°F.

**Rubber Processing Aids**

Natural and Synthetic Rubbers

"Castorwax" is highly compatible with synthetic and natural rubbers. It improves processing and reduces the nerve of the rubber with no or minimal loss in physical properties. It assists in the dispersion of dry materials and improves the extrusion rate with better knitting, imparting excellent shape retention on extrusion with less shrinkage of molded goods.

When used at a level of from 1.5 parts in the molding of butadiene-styrene into shoe soles and heels, "Castorwax" provides mold release and antiblocking with a dry feel. It imparts a luster to the finished product without blooming.

"Paricin" 220 at 1 to 2 parts PHR in molded rubber products gives good appearance and high sheen. Hydroxystearic Acid functions as an activator and internal lubricant for natural and synthetic rubbers.

Butyl Rubber

"Paricin" 1 functions as an internal lubricant at a level of from 2 to 5 parts PHR of halogenated or unhalogenated butyl polymers. At more than 5 to 10 parts PHR it exudes to the surface to form an invisible coating which markedly improves the calenderability and molding characteristics of the polymer. The coefficient of friction of the polymer is reduced over 58% for both the halogenated and the unhalogenated butyl polymer.

Ethylene-Propylene Rubber

"Paricin" 1 at a level of 4 parts PHR of press-cured, ethylene-propylene rubber completely eliminates surface tack.

---

**Self Emulsifying "Wax"**

| | |
|---|---|
| Potassium lauryl sulfate | 5 |
| Stearyl alcohol | 45 |
| Polyethylene glycol 400 | 4 |

---

**Investment Casting Wax**

| | No. 1 | No. 2 | No. 3 | No. 4 | No. 5 |
|---|---|---|---|---|---|
| Candelilla | 27 | | | | 31 |
| "Staybelite" ester 10 | 6 | — | — | — | — |
| "Piccovar" 420 | 20 | — | — | — | — |
| "Be Square" 170/175 Wax | 15 | 25 | — | 25 | 17 |
| "Castorwax" | 20 | 15 | 65 | 30 | 22 |
| "Alpco" 16 | — | 10 | — | — | — |
| "Piccopale" 100-SF | — | 30 | — | 30 | — |
| Burgundy pitch | — | 15 | 10 | 10 | 7 |

| | | | | | |
|---|---|---|---|---|---|
| "Aroclor" 4465 | — | — | 25 | — | — |
| "Aroclor" 5460 | — | — | — | — | 23 |

CHAPTER XVIII

# TEXTILE SPECIALTIES

**Durable Finish for Cotton and Rayon**

| | |
|---|---|
| "Gantrez" AN-139 | 34 |
| "Elvanol" 71-24 (polyvinyl alcohol, medium viscosity) | 62 |
| Sodium carbonate, anhydrous | 4 |
| Water | 1900 |

**Note:** Insolubilized by heat (150°C for 5 min).

---

**"Wash and Wear" Cotton Finish**

| | |
|---|---|
| "Acrite" 100 | 140 lb |
| Magnesium chloride, Hexahydrate | 28 lb |
| Nonionic wetting agent | 1 lb |
| Water to make a 100-gal mix | |

After padding, the fabric is dried, ideally to a moisture content of approximately 4 to 8%, and then cured. Optimum cure conditions are 1½ to 3 minutes at temperatures from 315 to 335°F. Higher temperatures will tend to discolor the fabric, whereas shorter times will not produce the desired degree of reaction and wash-wear performance. In the event that minor discolorations are encountered, a typical sodium perborate afterwash or a "topbluing" may be useful.

Following the cure step, an afterwash is recommended to eliminate any fabric odor due to trace quantities of unreacted product. It is suggested that the first wash box contain about 0.5 wt% aqueous sodium hydroxide solution to facilitate removal of any unreacted "Acrite". Depending on the wash capacity, it may also be advisable to neutralize the fabric with a small amount of acetic acid later in the washing train.

## Crease-Resistant Finish for Cotton

No. 1

| | |
|---|---|
| "Elvanol" 50-42 (5% solution) | 100.0 lb |
| "Paraplex" G-62 | 20.0 lb |
| "Eponite" 100 | 100.0 lb |
| 40% aqueous zinc fluoborate | 12.5 lb |
| | 100.0 gal mix |

No. 2

| | |
|---|---|
| "Elvanol" 50-42 (5% solution) | 100.0 lb |
| "Moropol" 700 | 20.0 lb |
| "Eponite" 100 | 75.0 lb |
| 50% ethylene urea-formaldehyde resin | 60.0 lb |
| 40% aqueous zinc fluoborate | 13.0 lb |
| | 100.0 gal mix |

The cotton goods should contain less than 0.1% alkal.

## Crease-Resistant Rayon

| | |
|---|---|
| "Elvanol' 50-42 (5% solution) | 100.0 lb |
| "Moropol" 700 | 16.0 lb |
| "Eponite" 100 | 62.0 lb |
| 50% ethylene urea-formaldehyde | 62.0 lb |
| 40% aqueous zinc fluoborate | 11.5 lb |
| | 100.0 gal mix |

## Fabric Softener

| | |
|---|---|
| "Miranol" SHD | 3.50 |
| Phosphoric acid | 0.04 |
| Water | 96.46 |

This may be perfumed and colored to suit.

## Laundry Goods Softener

(Commercial)

| | |
|---|---|
| "Varisoft" 222 | 54 |
| Water | 46 |

Heat water to 120 to 140°F and pour into the "Varisoft" 222 (making sure the softener concentrate is above 60°F). Stir, preferably with paddle type action, since at this stage there will be gel formation. If a thinner gel is desired, a few percent of isopropanol may be added.

## Flame Resistant Awning Coating

| | No. 1 | No. 2 |
|---|---|---|
| "Alloprene" •20 | 4.4 | 4.4 |
| "Cereclor" 42 | 6.6 | 6.6 |

| | | |
|---|---|---|
| Pentachlorophenol | 0.5 | 0.5 |
| Trichloroethylene | 73.7 | 72.2 |
| Antimony oxide | 6.5 | 6.5 |
| Zinc borate | 1.5 | 1.5 |
| Zinc carbonate | 6.0 | — |
| Paris white | — | 3.0 |
| China clay | — | 3.0 |
| Organic pigment | 0.5 | — |
| Organic pigment | — | 2.0 |
| Hydrogenated castor oil | 0.3 | 0.3 |

No. 3

| | |
|---|---|
| "Parlon" S20 | 17.5 |
| "Clorafin" 40 | 24.7 |
| Tricresyl phosphate | 3.9 |
| Copper naphthenate, 6% | 10.6 |
| Calcium carbonate | 11.0 |
| Titanium dioxide, rutile | 32.3 |

---

**Rainproofing Polyester-Cotton**

No. 1

1. Prepare finish
   - 4.0% Isopropanol
   - 1.2% "Phobotex" f/t/c (water repellent)
   - 0.3% Catalyst RB (water repellent catalyst)
   - 0.3% Acetic acid
   - 1.0% "Ceramine" HC (cationic softener)
   - 4.5% "Aerotex" 23 Special or (cross-linking resin)
   - 20.0% "Permafresh" Reactant 183 for deferred cure
   - 1.7% Catalyst X-4 (cross-linker catalyst)
   - 4.0% "Zepel" B (fluochemical oil/water repellent)
2. Pad-dry
3. Cure 1 to 3.5 min at 340 to 325°F
4. Sanforize

No. 2

1. Prepare finish with:
   - 0.2% "Synthrapol" KB (wetting agent)
   - 20.0% "Valrez" H-17 (cross-linking resin)
   - 3.0% "Atcosoft" PE (polyethylene softener)

3.0% "Poly-Tex" CL-304 (hand builder)
4.0% "Atco-pel" S-400 (silicone water repellent)
4.0% "Atcofix" SGA (water repellent catalyst)
4.0% "Valcat" 1700 (cross-linker catalyst)
0.5% citric acid

2. Pad-dry
3. Deferred cure in garment form.

No. 3

1. Prepare finish with:
   0.1% "Triton" X-155 (emulsifier)
   2.5% "Norane" F (fluorophilic reactant)
   20.0% "Fixapret" CP-40 (cross-linking resin)
   1.0% "Mykon" SF (polyethylene softener)
   40% Catalyst COT (cross-linker catalyst)
   2.5% FC-208 (fluorochemical oil/water repellent)
2. Pad-dry
3. Deferred cure in garment form

No. 4

| | |
|---|---|
| THPC | 250 g/l |
| NaOH 100% | 25 g/l |
| Urea | 50 g/l |
| Polyethylene emulsion | 20 g/l |
| Carbamate nonionic wetting agent | 1 g/l |
| Resin, dimethylolmethoxyethyl | 70 g/l |

The fabrics are padded at 80°F with this flame retardant finish, at 80% wet pickup, dried at 200°F, then cured for 3 min at 320°F. The fabrics are neutralized by padding at 80°F with a solution of:

| | |
|---|---|
| Ammonium sulfate | 60 g/l |
| Sodium bicarbonate | 30 g/l |
| Sodium carbonate | 36 g/l |

Left standing for 15 min, soaped at 180°F for another 15 min with:

| | on weight of fabric |
|---|---|
| Nonionic detergent | 1% |
| Sodium carbonate | 4% |
| Sodium perborate | 4% |

and then rinsed and dried.

No. 5

| | |
|---|---|
| "Pyrovatex" CP | 25-35 |
| "Aerotex" resin 23 special | 10.00 |
| Nonionic wetting agent | 0.05 |
| Urea | 1.00 |
| Polyethylene softener (30% solids) | 2.00 |
| Ammonium chloride | 0.40 |

Add: Prediluted nonionic wetting agent
Prediluted urea
Prediluted polyethylene softener
Prediluted ammonium chloride
Bring up to volume.

Pad bath temperature should not exceed 85°F. Generally, the wet pickup should be between 70 and 80%. Recommended curing conditions are 1 min 45 seconds at 350°F.

No. 5

| | |
|---|---|
| Chlorinated Paraffin | 362 |
| DiOctyl Phthalate | 362 |
| 2% Sodium carboxy methyl cellulose | 100 |
| Emulsifying agent | 16 |
| Water | 600 |
| Antimony oxide | 364 |
| Polyvinyl chloride latex | 1500 |
| Silicone antifoam | 1 |

The liquor is applied to the fabric by padding, and any surplus is expressed by adjustment of the nip to give a desired wet "add-on" which is a direct function of the final dry "add-on". The treated fabric is then dried, and cured at 120-140°C for 2-4 min. Washing and treatment with an appropriate softening agent may then be carried out if necessary in conventional plant used for this purpose.

In special circumstances the emulsion may be applied by brush or spray. For example, the hessian backing of a carpet can be flameproofed after manufacture by spraying or brushing the emulsion on to the reverse side, subsequently drying and curing. However applied, the emulsion is absorbed by the fibres, and thus a continuous surface coating of the fibre is formed.

**Flame-Water-Rot-Proof Textile Coating**

| | |
|---|---|
| "Alloprene" 20 | 5.50 |
| "Cereclor" 48 | 5.00 |
| Tricersyl phosphate | 3.50 |
| "Mobil" Wax 2305 | 1.50 |
| Pentachlorophenol | 0.50 |
| Solvent naphtha | 21.00 |
| Antimony oxide | 7.00 |
| Zinc borate | 5.00 |
| Paris white | 2.00 |
| China clay | 2.60 |
| "Polymon" Green GN500 | 0.40 |
| Cellulose ether | 0.75 |
| Sodium hexametaphosphate | 0.1 |
| Water | 45.65 |

The impregnant is prepared by warming together the "Mobil" Wax, "Cereclor" 42 and tricresyl phosphate to give a homogenous mixture which is added to the solvent. The chlorinated rubber and pentachlorophenol are dissolved in the resulting solution. All the pigments are then added and ball-milled for 2-3 hours. The mill-base is dispersed in the water phase using high speed stirring. A smooth cream is obtained of density to 9.0-9.5 lb/gal according to the pigment content. The emulsion is of good stability and can be diluted in water.

---

**Run-Resistant Hosiery Spray**

| | |
|---|---|
| PVP/VA E-355 | 4.0 |
| Methyl phthalate | 0.1 |
| S.D. 40 anhydrous ethanol | 6.0 |
| Propellant 11/12 (65/35) | 89.9 |

---

**Nylon Size**

| | |
|---|---|
| "Gantrez" AN-119 | 720 |
| "Gafanol" E-1500 | 160 |
| Water, distilled | 2960 |

---

**Binder for Vat Colors on Glass Fiber Fabrics**

| | |
|---|---|
| "Gantrez" AN-139 | 3.90 |
| Polyvinyl alcohol | 6.75 |
| Sodium carbonate | qs to pH 3.5-3.7 |

---

**Water-Soluble Vehicle for Oilbased Pigment Printing Pastes**

| | |
|---|---|
| "Gantrez" AN-169 | 1.0 |
| Mineral spirit (paraffin oil, kerosene, or distillate fuel oil) | 3.0 |
| Ethanolamine | 1.0 |
| Water | 95.0 |

---

**Fiber Glass Finish**

| | |
|---|---|
| "Gantrez" AN-139 | 0.5 |
| "Rezosol" 550 (20% aqueous dispersion) | 5.0 |
| Water | 94.5 |

**Note:** This finish is heat-set at 320°F. The coated fabric is then treated with 2% stearato chromic chloride in water and again heated at 320°F.

---

**Cotton Batting Cushions**

A water solution of any of the following is sprayed on the batting to saturation. Then it is squeezed and dried and cured.

20% Resin Solids

No. 1

| | |
|---|---|
| Methylated methylol melamine | 50% |
| Vinyl acetate | 50% |

Buffer formaldehyde acceptor catalyst (magnesium chloride) water.

No. 2

| | |
|---|---|
| Urea formaldehyde (modified) | 50% |
| Vinyl acrylate | 50% |

Buffer or formaldehyde acceptor catalyst (magnesium chloride) water.

No. 3

| | |
|---|---|
| Methylo. imidazolidone | 50% |
| Vinyl acetate-acrylate | 50% |

Catalyst (zinc chloride) water.

No. 4

| | |
|---|---|
| Methylated methylol melamine | 50% |
| Vinyl chloride-acetate | 35% |
| Styrene butadiene | 15% |

Buffer or formaldehyde acceptor catalyst (magnesium chloride) water.

---

**Nonwoven Fabric Binder**

Spray Bonding

No. 1

| | |
|---|---|
| "Poly-Tex" Emulsion (46% NV) | 20.0 lb |
| Water | 30.0 lb |
| Ammonium chloride, or | 50.0 g |
| Oxalic acid | 40.0 g |
| "Triton" X-100 | 50.0 g |

The webs are usually sprayed on both sides either as they emerge from the crosslapper or by spraying one side, running the web through the oven and spraying the other side after the web turns over a roll for the return trip through the oven. A combined dry and cure time of 3 min at 325°F is desirable, but but both lower and higher temperatures have been found satisfactory with some time adjustment.

Saturation Bonding
No. 2

| | |
|---|---|
| "Poly-Tex" Emulsion (46% NV) | 72.0 lb |
| Water | 144.0 lb |
| Ammonium chloride or oxalic acid | 4.5 oz |
| "Triton" X-100 | 4.5 oz |
| "Dow Corning" Antifoam A[2] | 500.0 cc |

The drying and curing requirements will be generally the same as used for the spray-bonded web. In either case where a nonreactive emulsion is used, the catalyst may be omitted or where a cellulosic fiber is involed, the use of a small amount (5-15%) of a catalyzed urea or melamine resin such as Aerotex M-3 may prove desirable.

As might be expected, spray bonding is normally used for the loftier webs, commonly called fiber fill, and saturation bonding is used for denser webs such as apparel interliners, sheets, gowns, etc.

Chapter XIX

# MISCELLANEOUS

## Artificial Perspiration

| | |
|---|---|
| Water | 840 |
| Sodium chloride | 50 |
| Glacial acetic acid | 50 |
| Butyl acid | 30 |
| Isovaleric acid | 30 |

## Microbial Growth Inhibitor

p-hydroxybenzoates

| | Required for Inhibition, % | | | |
|---|---|---|---|---|
| Microorganisms | Methyl | Ethyl | Propyl | Butyl |
| Aspergillus niger ATCC 1025-1 | 0.100 | 0.040 | 0.0200 | 0.0200 |
| Penicillium digitatum ATCC 10030 | 0.050 | 0.025 | 0.0063 | <0.0032 |
| Rhizopus nigricans ATCC 6227A | 0.050 | 0.025 | 0.0125 | 0.0063 |
| Trichoderma lignorum ATCC 8678 | 0.025 | 0.013 | 0.0125 | 0.0063 |
| Chactomium globosum ATCC 6205 | 0.050 | 0.025 | 0.0063 | <0.0032 |
| Trichophyton mentagrophytes ATCC 9533 | 0.016 | 0.008 | 0.0040 | 0.0020 |
| Trichophyton rubrum ATCC 10218 | 0.016 | 0.008 | 0.0040 | 0.0020 |
| Candida albicans ATCC 10231 | 0.100 | 0.100 | 0.0125 | 0.0125 |
| Saccharomyces cerevisiae ATCC 9763 | 0.100 | 0.050 | 0.0125 | 0.0063 |
| Saccharomyces pastorianus ATCC 2366 | 0.100 | 0.050 | 0.0125 | 0.0063 |

| | | | | |
|---|---|---|---|---|
| Bacillus subtilis | | | | |
| ATCC 6633 | 0.200 | 0.100 | 0.0250 | 0.0125 |
| Bacillus cercus var. mycoides | | | | |
| ATCC 6462 | 0.200 | 0.100 | 0.0125 | 0.0063 |
| Staphylococcus aureus (Micrococcus pyogenes var. aureus) | | | | |
| ATCC 6538P | 0.400 | 0.100 | 0.0500 | 0.0125 |
| Sarcina lutea | 0.400 | 0.100 | 0.0500 | 0.0125 |
| Klebsiella pneumoniae | | | | |
| ATCC 10031 | 0.100 | 0.050 | 0.0250 | 0.0125 |
| Escherichia coil | | | | |
| ATCC 9637 | 0.200 | 0.100 | 0.1000 | 0.4000 |
| Salmonella typhosa | 0.200 | 0.100 | 0.1000 | 0.1000 |
| Salmonella shollmuelleri | 0.200 | 0.100 | 0.0500 | 0.1000 |
| Proteus vulgaris | | | | |
| ATCC 8427 | 0.200 | 0.100 | 0.0500 | 0.0500 |
| Aerobacter aerogenes | | | | |
| ATCC 8308 | 0.200 | 0.100 | 0.1000 | 0.4000 |

---

### Noncaking Powders

No. 1

The addition of less than 1% "Aerosil" R-972 to a hygroscopic powder makes it noncaking and free-flowing.

No. 2

10% "CAB-O-SIL" can be used to prevent DDT from caking and 0.4% has produced a free-flowing sulphur. 0.5 to 1% "CAB-O-SIL" is usually quite effective in baby powders, foot powders and all caky materials.

---

### Dustless Powders

As little as 1% of "Pluronic" L61 or L62 eliminates dusting problems encountered with finely divided talc. In tablets or depilatory sticks, "Pluronic" polyols, notably F68, can be used as a binder or plasticizer and in tablet coating, although other Pluronic flaked solids could also be used.

---

### Dust Control Oil

| | |
|---|---|
| Mineral oil | 96 |
| "Variquat" K300 | 2 |
| "Varonic" L202 | 2 |

The composition also exhibits a degree of germicidal and fungicidel activity because of the biocidal properties of the "Variquat K300."

### Thickening of Oils

Weight percent "Thixcin" R to produce ointment consistency shown

| | Liquid | Semi-solid | Cream | Soft | Stiff |
|---|---|---|---|---|---|
| Almond oil, expressed | 1% | — | 3% | 5% | 7% |
| Castor oil | 1 | — | 3 | — | 5 |
| Cottonsed oil | 1 | 3 | 5 | — | 7 |
| Linseed oil | 1 | 3 | 5 | — | 7 |
| Petrolatum, heavy liquid | 3 | 5 | 7 | 9 | 10 |
| Polyethylene glycol 400 | 1 | 3 | 5 | 7 | — |

"Thixicin" R is effective as a thickening and suspending agent in a variety of organic liquids such as: Alcohols, glycols, glycol ethers, esters, ketones, mineral oils, polyglycols, and vegetable oils. "Thixcin R" is not suitable for use in aromatic hydrocarbons or water alone.

In emulsion systems "Thixcin" R is dispersed in the oil phase before emulsification. Both oil-in-water and water-in-oil emulsions can be formed. Because "Thixcin" R remains in the oil phase, it has relatively little effect on the viscosity of oil-in-water systems. However, water-in-oil emulsions containing "Thixcin" R are thickened considerably.

High shear is essential for good dispersion and thixotropic efficiency. "Thixcin" R can be dispersed in laboratory blenders, colloid mills, ointment mills, Eppenbach "Homo-Mixers", or other standard high-shear equipment.

In dispersing "Thixcin" R the temperature should be maintained within the range 90 to 130°F.

---

### Thickening Aqueous Systems

(Using 1% Bipolymer XB-23)

Sols of "XB-23" are prepared by dispersing the biopolymer in aqueous medium at ambient (25°C) temperature using mechanical stirring. Agitation speed should be controlled to avoid entraining air into the sol. The speed of the stirrer is adjusted so that the vortex reaches half the depth of the aqueous medium. "XB-23" is then added into the vortex and stirring is continued until the "XB-23" hydrates enough to remain in suspension. In water, 30 sec of stirring is sufficient but in acid and salt solutions, two minutes or longer may be necessary. Complete hydration and attainment of maximum viscosity occur without further agitation.

| System | Viscosity* 24 hr | 48 hr | 72 hr |
|---|---|---|---|
| $H_2O$ | 3450 | 3500 | 3300 |
| 36% NaCl | 3950 | 4450 | 4750 |
| 5% NaCl | 3350 | 3700 | 3800 |
| 10% KCl | 3650 | 3750 | 3800 |
| 5% $MgSO_4$ | 3400 | 3800 | 3900 |
| 8% NaCl+2.5% $CaCl_2$ | 2050 | 2450 | 2650 |
| 15% HCl | 2750 | 2600 | 2500 |
| 60% Acetic acid | 1900 | 2550 | 2900 |
| 15% $H_3PO_4$ | 2850 | 3100 | 3100 |
| 5% HF | 2000 | 2400 | 2500 |
| 5% Sodium lauryl sulfate | 3900 | 3650 | 3650 |
| 5% Ethoxylated phenol (nonionic) | 3600 | 3500 | 3400 |

*Brookfield RVF No. 4 Spindle—20 RPM

**Gels**

The following percentages of "CAB-O-SIL" by weight are required to produce soft gels with the corresponding liquids, listed in order of decreasing polarity:

12% in water
11% in ethylene glycol
9% in butyl alcohol
8% in turpentine
7% in benzene

**Suspending Agent**

The shelf stability of organic and water-base suspensions can be improved with the use of 0.2 to 0.3% "CAB-O-SIL", which retards the settling of high density pigments. Flow properties of such systems are also simultaneously controlled.

**Thixotropic Agent**

The addition of 1% "CAB-O-SIL" to a nonthixotropic mineral oil has increased the viscosity from 410 centipoises (Brookfield viscosity) to 8,000 centipoises at 2 r.p.m. and 3,500 centipoises at 20 r.p.m. "CAB-O-SIL" can be used effectively in salves and ointments where other pigments are present.

**Hydraulic Fluid**

| | |
|---|---|
| 4-Phenylmorpholine | 3.00 |
| Sodium tetraborate pentahydrate | 0.60 |
| "Gafcol®" EM | 6.06 |
| Gantrez" M-094 | 2.00 |
| Diethylene glycol diethyl ether | 10.00 |
| Polymethoxy-l-alkanols (3:1) | 81.04 |

**Water Filtering Membrane**

| | |
|---|---|
| Cellulose acetate E-398-3) | 25 |
| Formamide | 30 |
| Acetone | 45 |

Casting of membranes from this type of solution, as well as the subsequent leaching, is carried out at room temperature. Because of this, highly reproducible membranes can be cast. These membranes have permselectivity with the added advantage of higher throughput rates. Room temperature casting also adds economy to the cost of membrane fabrication.

---

**Chemiluminescent Substance**

U.S.P. 3 315,176

| | |
|---|---|
| Paraffin | 15-55 |
| Microcrystalline wax | 15-55 |
| Mineral oil | 5-25 |
| Methylphenylsilicone | 1-20 |
| Tetrakis (dimethylamino) ethylene | 20-30 |

**Modelling Clay**

| | |
|---|---|
| Sulfur, powdered | 66 |
| Kaolin | 400 |
| Lanolin | 120 |
| Glycerin | 80 |

---

**Lead Sulfide Mirror**

| | |
|---|---|
| 1. Lead acetate | 1 |
| Water | 24 |
| 2. Caustic soda | 4 |
| Water | 32 |
| 3. Thiourea | 1 |
| Water | 48 |

Four parts of Solution 3 and 1 part of Solution 1 are mixed, after which one part of Solution 2 is added. This is applied immediately to the cleaned surface.

---

**Bottle Scratch Preventor**

During travel on conveyer belt, from blower which hot, glass bottles develop scratches from joggling. These are prevented by spraying with

| | |
|---|---|
| "Myrj" 52-S | 0.02-0.1 |
| Water, to make | 100 |

# APPENDIX

## Federal Laws Regulating Foods, Drugs and Cosmetics

Anyone who plans to market a food, drug cosmetic, or chemical specialty should be thoroughly familiar with federal laws and regulations pertaining to that particular product.

The Federal Food and Drug Law, passed in 1938, was comparatively lenient in its restrictions. But discovery of more potent drugs, new food additives, and powerful insecticides posed safety problems. In some instances dangerous toxicity of a new compound was not suspected prior to marketing. Then, when its use was widespread, serious poisonings and other deleterious actions were reported.

In 1962 every restrictive amendments to the 1938 law were passed, demanding far more rigid proof of the safety of any drug or chemical used on or by humans, and also covered products used for pet animals as well as food animals.

Responsibility is still divided among three government agencies. The Federal Trade Commission is charged with the responsibility of policing advertising to check false, misleading or illegal advertising statements. One exception is drug advertising to physicians, which comes under FDA jurisdiction.

The Department of Agriculture has jurisdiction over pesticides and their labeling and additives to foods for animals being raised for human consumption.

The Food and Drug Administration (FDA), a bureau of the Department of Health, Education and Welfare, has the broadcast regulatory powers. The FDA controls the marketing of all foods, medications and cosmetic products for human use and all veterinary medications.

### *CHEMICAL SPECIALTIES*

Household cleaners, spot removers, polishes, floor and furniture waxes, bleachers and washing fluids as well as industrial chemicals are in a "twilight zone" as far as regulation is concerned.

Many of these substances are toxic and potentially dangerous, yet there are no nontoxic substitutes for many of them. Farm chemicals are under control of the Department of Agriculture, but household chemical specialties come under surveilance of the FDA.

Special precautionary labeling is required, and on some, an antidote must be printed on the label, so that effective corrective measures can be taken promptly in case of accidental ingestion.

As in the case of drugs or cosmetics, any manufacturer contemplating marketing any household product should check with FDA concerning restrictions and mandatory labeling. These restrictions are generally quite liberal and are designed only to protect the public. Accurate, official information is so readily obtained that there is no excuse for anyone to offer dangerous products for sale without the proper safeguards. To neglect this means not only official censure, but could weigh heavily against the manufacturer if he should become the defendant in a law suit for damages caused by his product.

### *DEVICES*

This is a general term covering any device either mechanical, electrical, or hydraulic, that purports to cure or alleviate any physiological ailment. Massage vibrators, "whirpool" baths, heat lamps, diagnostic instruments, ultra violet lamps are all subject to FDA regulations if the device is to be sold to the public and if some curative claim is made for it. Obviously "magnetic belts" that claim to cure all kinds of diseases would never get official permission to make such claims. But there are some devices, especially those used with specific medications (for instance, a special vaporizer for inhalation therapy) that have significant value and FDA will approve thoughtful and *proven* claims for these devices.

### *COSMETICS*

Cosmetics are not restricted as stringently as drugs, but the definition of a cosmetic is very narrow. Any preparation used *only* for cleansing or beautification of the skin, hair, or finger nails is considered a cosmetic. There are certain prohibited drugs or chemicals which are considered too dangerous for use in cosmetics, but on the whole it is up to the manufacturer to market products that are safe. Cosmetic claims can be quite flamboyant, as long as the claims are only for beautification. However, any claim of a medicinal nature, even if only implied, immediately places the product in the drug class, which means that official permission to market the product is mandatory.

The law does not forbid the use of poisonous substances in a cosmetic. The ban applies only when the quantity of a poisonous ingredient *may* be injurious to users, *under the directions of use as given by the manufacturer*. The product may be extremely hazardous if misused, yet still be acceptable, under the law, *provided* safe directions are included and proper caution statements are made on the label.

While the Federal Trade Commission has charge of all cosmetic *advertising*, it is the FDA that controls *labeling*. The FDA has the authority to seize any product in interstate commerce which it considers misbranded. The term "labeling" is a rather all-inclusive designation and covers much more than just the label pasted on the container. Any written, printed or graphic material accompanying the package is considered part of labeling, even if it is delivered by mail separately from the actual product. A display in a store, together with the product, is also "labeling"; so is any literature handed out by demonstrators.

The manufacturer of a cosmetic is literally walking a "verbal tight rope", as the following *partial* list of taboos illustrates. Inclusion of any of the following designations are sufficient to bring a charge of misbranding a cosmetic, circulating cream, contour cream, deep pore cleanser, enlarged pore paste, depilatories for *permanent* removal of hair, eyelash grower, eye wrinkle cream, hair color restorer, hair grower, hair restorer or revitalizing preparation, muscle oil, nail grower, nonallergic products, peroxide cream, scalp food, skin conditioner or tonic, skin food or nourishing cream, skin texture preparations, stimulating creams, tissue cream, wrinkle and double chin eradicator, any representation that a product is superior in cosmetic value because it contains a vitamin.

The law does not require that a cosmetic must contain a list of its ingredients, but if the labeling puts it in the "drug" class, it must list all ingredients with the quantities. Any statement that a cosmetic will cure or alleviate dandruff, will promote a "healthy skin or scalp", alleviate itching, revitalize dry skin, cure acne, eczema, freckles, rash or any "skin trouble", places the product in the "drug" class.

The manufacturer or distributor of a cosmetic has the full responsibility for a truthfully labeled cosmetic, as defined by FDA. The FDA is authorized to approve the label or labeling of a cosmetic, but it usually will give an informal opinion as to whether a proposed label is to be construed in fact as a drug label. But even such informal opinion does not relieve the manufacturer of his responsibility, because such opinion is not an official acceptance of the product nor its labeling.

### *DRUGS*

Any product, whether used externally or internally, that purports to treat any abnormal human condition is classed as a drug; the definition is broad, and applies to products which can be considered preventive rather than curative or therapeutic.

Drugs are divided into two broad classes. The so-called "over the counter" drugs are considered reasonably safe (and effective) to be used by the general public, without the recommendation and supervision of a physician, and can be purchased without a prescription.

The regulation of these drugs is very strict, specifying how directions should be stated, warnings and cautions to be prominently displayed on the label, and an accurate list of the ingredients, with quantities given for the active constituents. Also, curative claims are severely restricted. For instance, an aspirin containing tablet cannot claim to relieve pain of rheumatism; it must state "for the temporary relief of minor aches and pains of rheumatism", and in addition must carry the accepted warnings about the possibility of adverse reactions, and the precaution to keep out of reach of children!

All labels, claims and opposed advertising should be submitted to the FDA before any "over the counter" drug is marketed.

### *PRESCRIPTION DRUGS*

Drugs that are considered potentially dangerous for self medication can be dispensed only by licensed pharmacists (or hospital dispensaries), on a physician's prescription, or dispensed by the physician to his own patients.

Prior to 1962, drugs were cleared for sale by proving safety when used according to directions. There was no legal basis for withholding permission to market a drug because of exaggerated or false statements of curative claims. The 1962 drug law amendments change this. Now *all* drugs, whether prescription or nonprescription, must show acceptable proof that they are not only safe, but will have the therapeutic effect claimed.

Drugs in both classes that had been approved for safety prior to the 1962 law must now have acceptable clinical proof that they are effective. If such proof cannot be presented, the drugs are ordered off the market, despite the fact that they may have been sold for the past, since 1938! Many an old time favorite has had to drastically revise claims and often the formula, or else face economic extinction.

The FDA has the final decision as to whether a drug can be sold without a prescription or not. The manufacturer cannot designate the catagory.

*NEW DRUGS*

All new drugs must of course comply with the 1962 law. This often presents a formidable obstacle. Before trying a new drug on humans, the FDA must be petitioned for permission. Unless there is incontravertible evidence of safety, FDA will not approve an Investigational Drug Application (IDA). This means, in the majority of cases, elaborate, carefully controled experimentation on laboratory animals. Usually tests must be done with several species of animals, under supervision of qualified pharmacologists.

If FDA is satisfied with the animal tests, it may consider the IDA favorably. Qualified physicians (usually specialists in their fields) with adequate hospital affiliations, are permitted to administer the new drug to humans. This "clinical investigation" is hedged in by strict rules of reporting progress to FDA. The clinical testing may take three months up to several years, and may require as few as 20 patients or may have to be tried on several thousand patients.

When all the clinical work is finished, the complete history of the project, along with the reports of the clinical investigators is then submitted to the FDA, as part of a New Drug Application (NDA). If government experts are satisfied with the report, the NDA is validated and the drug can be offered for sale.

But the FDA continues its surveillance and any violation of approved label claims can bring revocation of the NDA thereby terminating the sale of the drug. Any change, no matter how slight, in the formula, or method of manufacture requires sanction by FDA; unauthorized changes invalidate the NDA.

An interesting side light is that all advertising material for a prescription type drug is considered part of labeling. An advertisement in a medical journal, a display at a medical convention, or a brochure or booklet mailed to a physician, are all considered as much a part of a label as the piece of paper pasted on the drug container.

Another pitfall for the uninitiated is the definition of a "new drug". FDA contention is that combinations of drugs that are in proportions, or identity, different than those acknowledged as "safe and effective", are to be treated as "new drugs". This holds true even if every *single* drug in such a new combination has been used for years and has been approved. For instance, aspirin is a common drug, fully approved; so is vitamin C, and so is caffein. Yet, if these well known constituents are combined, and especially if offered in novel proportions and in a new vehicle, or form, such as a liquid solution, or an effervescent tablet, they are a "new drug" and must go through the formality of obtaining an approved NDA.

Reading the provisions of the law as passed by Congress in 1962 is not too helpful to a manufacturer. There have been so many supplemental regulations and new rulings (they come with great regularity and rapidity), that any book purporting to give *all* interpretations of the law would be critically obsolete before it could be printed.

There is one other important provision that must be observed. Any person or company that manufactures drugs or cosmetics must permit inspectors of the FDA free access to the laboratories and manufacturing premises "during reasonable working hours". At least an annual inspection is supposed to be made to ascertain if facilities are adequate and if properly qualified personnel is in charge of the manufacturing and assay of raw materials and finished products. This applies as well to any laboratory engaged in assay for other manufacturers, or any plant making products for distributors, even if the laboratory is not engaged in interstate commerce.

Faced with all these strict regulations, it is only prudent for any manufacturer to check with the FDA on what will be required, before any serious committment is made on any new product, be it in the drug, food or cosmetic field. The Food and Drug Administration will give a reasonably prompt answer to any informal inquiry, and will in most cases indicate what will be needed for a formal application, for a proposed drug or formula. All informal inquiries should be addressed to Food and Drug Administration, Arlington, Virginia. Household chemical specialties, and insecticides inquiries should be addressed to the Department of Agriculture, Washington, D.C.

# TABLES

## Weights and Measures

### Troy Weight

24 grains = 1 pwt.
20 pwts. = 1 ounce
12 ounces = 1 pound

### Apothecaries' Weight

20 grains = 1 scruple
3 scruples = 1 dram
8 drams = 1 ounce
12 ounces = 1 pound

The ounce and pound are the same as in Troy Weight.

### Avoirdupois Weight

27 11/32 grains = 1 dram
16 drams = 1 ounce
16 ounces = 1 pound
2000 lbs. = 1 short ton
2240 lbs. = 1 long ton

### Dry Measure

2 pints = 1 quart
8 quarts = 1 peck
4 pecks = 1 bushel
36 bushels = 1 chaldron

### Liquid Measure

4 gills = 1 pint
2 pints = 1 quart
4 quarts = 1 gallon
31½ gals. = 1 barrel
2 barrels = 1 hogshead
1 teaspoonful = ⅛ oz.
1 tablespoonful = ½ oz.
16 fluid oz. = 1 pint

### Circular Measure

60 seconds = 1 minute
60 minutes = 1 degree
360 degrees = 1 circle

### Long Measure

12 inches = 1 foot
3 feet = 1 yard
5½ yards = 1 rod
5280 feet = 1 stat. mile
320 rods = 1 stat. mile

### Square Measure

144 sq. in. = 1 sq. ft.
9 sq. ft. = 1 sq. yard
30¼ sq. yds. = 1 sq. rod
43,560 sq. ft. = 1 acre
40 sq. rods = 1 rood
4 roods = 1 acre
640 acres = 1 sq. mile

### Metric Equivalents

#### Length

1 inch = 2.54 centimeters
1 foot = 0.305 meter
1 yard = 0.914 meter
1 mile = 1.609 kilometers
1 centimeter = 0.394 in.
1 meter = 3.281 ft.
1 meter = 1.094 yd.
1 kilometer = 0.621 mile

#### Capacity

1 U. S. fluid oz. = 29.573 milliliters
1 U. S. Liquid qt. = 0.946 liter
1 U. S. dry qt. = 1.101 liters
1 U. S. gallon = 3.785 liters
1 U. S. bushel = 0.3524 hectoliter
1 cu. in. = 16.4 cu. centimeters
1 milliliter = 0.034 U. S. fluid ounce
1 liter = 1.057 U. S. liquid qt.
1 liter = 0.908 U. S. dry qt.
1 liter = 0.264 U. S. gallon
1 hectoliter = 2.838 U. S. bu.
1 cu. centimeter = .061 cu. in.
1 liter = 1000 milliliters or 100 cu. c.

#### Weight

1 grain = 0.065 gram
1 apoth. scruple = 1.296 grams
1 av. oz. = 28.350 grams
1 troy oz. = 31.103 grams
1 av. lb. = 0.454 kilogram
1 troy lb. = 0.373 kilogram
1 gram = 15.432 grains
1 gram = 0.772 apoth. scruple
1 gram = 0.035 av. oz.
1 gram = 0.032 troy oz.
1 kilogram = 2.205 av. lbs.
1 kilogram = 2.679 troy lbs.

# Trade-Mark Chemicals

## Where to Buy Them

Numbers to the right of each item refer to the suppliers who are given in the list of sellers directly after this list.

Chemicals not sold under a trade mark may be located in the annual *Buyers' Guide* published by Chemical Week, 330 W. 42 St., New York, N.Y., 10036, and in the Green Book published by the *Oil, Paint and Drug Reporter*, 100 Church St., New York City, 10007.

# APPENDIX CHEMICALS (TRADEMARK)

# LIST OF SUPPLIERS

| *No.* | *Name* | *Address* |
|---|---|---|
| 1. | Advance Division | New Brunswick, N. J. |
| 3. | Airco Chem. & Plastics Div. | New York, N. Y. |
| 4. | Alco Chemical Corp. | Philadelphia, Pa. |
| 5. | Alcolac Chem. Corp. | Baltimore, Md. |
| 7. | Allied Chem. Corp. | New York, N. Y. |
| 9. | Amer. Cholesterol Prod. Inc. | Edison, N. J. |
| 10. | American Colloid Co. | Skokie, Ill. |
| 11. | Amer. Cyanamid Co. | Wayne, N. J. |
| 13. | Amer. Lignite Prod. Co. | Ione, Cal. |
| 15. | Amer. Litho. Varnish Co. | Kearny,N . J. |
| 17. | Amer. Maize Prod. Co. | New York, N. Y. |
| 19. | Amsco Solvents Inc. | Cincinnati, O. |
| 21. | ARCO Oil Corp. | Oklahoma City, Okla. |
| 23. | Argus Chemical Corp. | Brooklyn, N. Y. |
| 25. | Arizona Chem. Co. | New York, N. Y. |
| 27. | Armour Ind. Chem. Co. | Chicago, Ill. |
| 29. | Arkland Chem. Co. | Columbus, O. |
| 31. | Atlantic Richfield Co. | Philadelphia, Pa. |
| 33. | Atlas Chem. Ind. Inc. | Wilmington, Del. |
| 35. | Baird Chem. Ind. Inc. | New York, N. Y. |
| 37. | Baker Castor Oil Co. | Bayonne, N. J. |
| 39. | Balab | Burlingame, Cal. |
| 41. | Bareco Division | Tulsa, Okla. |
| 43. | BASF Corp. | New York, N. Y. |
| 45. | Basic Chemicals Division | Cleveland, O. |
| 47. | Borden Chemical Co. | New York, N. Y. |
| 49. | Brown Co. | New York, N. Y. |
| 50. | Buckman Labs. Inc. | Memphis, Tenn. |
| 51. | Cabot Corp. | Boston, Mass. |
| 53. | Campbell Son's Corp. H. T. | Baltimore, Md. |
| 54. | Carbon Dispersions Inc. | Newark, N. J. |
| 55. | Celanese Resins | Louisville, Ky. |
| 57. | Ciba Chem. & Dye Co. | Fairlawn, N. J. |
| 59. | Columbian Carbon Co. | New York, N. Y. |

61. Concord Chem. Co. ....................Camden, N. J.
63. Continental Oil Co. ....................New York, N. Y.
65. Croda Inc. ....................New York, N. Y.
67. Degussa Inc. ....................New York, N. Y.
69. Dewey & Almy ....................Cambridge, Mass.
71. Diamond-Shamrock Corp. ....................Cleveland, O.
72. Dispergent Co. ....................Guilford, Conn.
73. Dow Chem. Co. ....................Midland, Mich.
74. Dow-Corning Corp. ....................Midland, Mich.
75. Dragoco Inc. ....................Totowa, N. J.
77. Drew Chemical Corp. ....................New York, N. Y.
79. Du Pont de Nemours & Co. Inc., E. I. ....................Wilmington, Del.
81. Dura Commodities Corp. ....................New York, N. Y.
83. Durez Division ....................N. Tonawanda, N. Y.
85. Eastman Chem. Prod. Inc. ....................Kingston, Tenn.
87. Enjay Chemical Co. ....................New York, N. Y.
89. Enzyme Development Corp. ....................New York, N. Y.
91. Firestone Plastics Co. ....................Pottstown, Pa.
93. FMC Corp. ....................New York, N. Y.
94. Freeman Chem. Corp. ....................Port Washington, Wis.
95. GAF Corp. ....................New York, N. Y.
97. Geigy Chemical Corp. ....................Ardsley, N. Y.
99. General Chemical Co. ....................New York, N. Y.
100. General Electric Co. ....................Pittsfield, Mass.
101. General Mills Inc. ....................Kankakee, Ill.
103. General Tire-Rubber Co. ....................Akron, O.
105. Glyco Chemicals Inc. ....................New York, N. Y.
107. Goldschmidt Chem. Division ....................New York, N. Y.
109. Goodrich-Gulf Chem. Inc. ....................Cleveland, O.
110. Goodyear Tire-Rubber Co. ....................Akron, O.
111. Halby ....................Wilmington, Del.
113. Hampshire Chem. Division ....................Nashua, N. H.
115. Hercules Inc. ....................Wilmington, Del.
116. Heyden Division ....................New York, N. Y.
117. Holland-Suco Color Co. ....................Holland, Mich.
119. Houghton & Co., E. F. ....................Philadelphia, Pa.
121. Huber Corp., J. M. ....................Edison, N. J.
123. Humble Oil-Refining Co. ....................New York, N. Y.
125. I. C. I. America Inc. ....................Stanford, Conn.
127. International Talc Co. ....................New York, N. Y.
129. Jefferson Chemical Co. ....................New York, N. Y.
131. Johns-Manville ....................New York, N. Y.
133. Johnson & Son Inc., S. C. ....................Racine, Wis.
134. Kelco Co. ....................Clark, N. J.
135. Koppers Co. ....................Pittsburgh, Pa.
137. Lignosol Chemical Co. ....................Quebec, Ont., Canada
139. Lilly Co. Eli ....................Indianapolis, Ind.
141. Lubrizol Corp. ....................Cleveland, O.
142. 3M Co. ....................St. Paul, Minn.
143. Magie Bros. Oil Co. ....................Franklin Park, Ill.
145. Malmstrom Chem. Corp. ....................Linden, N. J.
147. Marine Colloids Inc. ....................Springfield, N. J.
151. McLaughlin Gormley King Co. ....................Minneapolis, Minn.

153. Metro-Atlantic Inc. ....Centredale, R. I.
155. Millmaster-Onyx ....New York, N. Y.
157. Minerals & Chem. Division ....Edison, N. J.
159. Miranol Chemical Co. ....Irvington, N. J.
161. Mobay Chemical Co. ....Pittsburgh, Pa.
163. Mona Industries Inc. ....Paterson, N. J.
165. Monsanto Chem. Co. ....St. Louis, Mo.
167. Morton Chemical Co. ....Chicago, Ill.
169. Naftone ....New York, N. Y.
171. National Lead Co. ....New York, N. Y.
173. National Starch & Chem. Corp. ....New York, N. Y.
175. Nease Chemical Co. Inc. ....State College, Pa.
177. Neville Chemical Co. ....Pittsburgh, Pa.
179. New Jersey Zinc Co. ....New York, N. Y.
181. Newport Division ....New York, N. Y.
183. Nopco Chem. Division ....Newark, N. J.
185. Norton Co. ....Worcester, Mass.
187. Nuodex Division ....Piscataway, N. J.
189. Olin Chemicals ....New York, N. Y.
190. Onyx Chemical Co. ....Jersey City, N. J.
191. Penn. Glass Sand Corp. ....Englewood Cliffs, N. J.
192. Penn. Industrial Chem. Corp. ....Clairton, Pa.
192A. Pennsalt Chemicals Corp. ....Philadelphia, Pa.
192B. Petrochemicals Co. ....New York, N. Y.
193. Petroleum Specialties Inc. ....New York, N. Y.
195. Pfizer & Co., Chas. ....New York, N. Y.
197. Phila. Quartz Co. ....Philadelphia, Pa.
199. Pilot Chemical Co. ....Santa Fe Springs, Cal.
201. PPG Industries Inc. ....Pittsburgh, Pa.
203. Proctor & Gamble Co. ....Cincinnati, O.
205. Radiant Color Co. ....Richmond, Cal.
207. Reichhold Chemicals Inc. ....White Plains, N. Y.
209. Richardson Co. ....Melrose Park, Ill.
211. R-I-T-A Chemical Co. ....Chicago, Ill.
213. Robeco Chemicals Inc. ....New York, N. Y.
215. Robertet Inc. P. ....New York, N. Y.
217. Rohm & Haas ....Philadelphia, Pa.
218. Ross-Rowe Inc. ....New York, N. Y.
218A. Rozilda Chemicals Inc. ....Paterson, N. J.
219. Scher Bros. ....Clifton, N. J.
221. Schimmel Co. ....New York, N. Y.
223. Shanco Plastics & Chem. Inc. ....Tonawanda, N. Y.
225. Shawinigan Prod. Corp. ....Englewood Cliffs, N. J.
227. Shell Chemical Co. ....New York, N. Y.
229. Sherwin Williams Co. ....Cleveland, O.
231. Sinclair Petrochemicals Inc. ....New York, N. Y.
232. Sindar Corp. ....New York, N. Y.
233. Sonneborn Division ....New York, N. Y.
235. Southern Clay & Co. ....Cassopolis, Mich.
237. Spencer-Kellog Division ....Buffalo, N. Y.
239. Standard Chem. Prod. Co. ....Hoboken, N. J.
241. Standard Oil Co. (N. Y.) ....New York, N. Y.
243. Stauffer Chem. Co. ....New York, N. Y.

245. Stepan Chemical Co. ..................Northfield, Ill.
247. Sun Chemical Corp. ..................Harrison, N. J.
247A. Sun Oil Co. ..................Philadelphia, Pa.
248. Swift & Co. ..................Chicago, Ill.
249. Tamms Industries Co. ..................Lyons, Ill.
251. Texas-US Chemical Co. ..................New York, N. Y.
253. Thiokol Chemical Corp. ..................Camden, N. J.
255. Thompson Weinman & Co. ..................Cartersville, Ga.
257. Titanium Pigment Corp. ..................New York, N. Y.
259. Troysan Chemical Co. ..................Newark, N. J.
261. UBS Chemical Co. ..................Cambridge, Mass.
263. Union Carbide Corp. ..................New York, N. Y.
265. Uni Royal Chem. Division ..................Naugatuck, Conn.
267. U. S. Borax & Chem. Corp. ..................Los Angeles, Cal.
269. Valchem ..................New York, N. Y.
271. Vanderbilt Co., R. T. ..................New York, N. Y.
273. Wallace-Tiernan Inc. ..................Belleville, N. J.
275. Washine Chemical Corp. ..................Lodi, N. J.
277. Witco Chemical Co. ..................New York, N. Y.
279. Wyandotte Chem. Corp. ..................Wyandotte, Mich.

# INDEX

# INDEX